HARCOURT

Math

Practice
Workbook

Grade 3

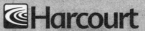

Harcourt

Orlando   Austin   Chicago   New York   Toronto   London   San Diego

Visit *The Learning Site!*
www.harcourtschool.com

ISBN 0-15-336475-0

10 11 12 13 14 15   018   10 09 08 07 06 05

CONTENTS

Algebra: Fact Families

Complete.

1. $7 + 4 = 11$, so $11 - 7 =$ ____

2. $16 - 7 = 9$, so $9 +$ ____ $= 16$

3. $6 + 8 = 14$, so $14 -$ ____ $= 6$

4. $12 - 6 = 6$, so ____ $+ 6 = 12$

5. $15 - 9 = 6$, so ____ $+$ ____ $= 15$

6. $7 + 7 = 14$, so $14 -$ ____ $=$ ____

Write the fact family for each set of numbers.

7. 6, 7, 13

_____

_____

_____

_____

8. 9, 9, 18

_____

_____

_____

_____

9. 3, 9, 12

_____

_____

_____

_____

10. 7, 8, 15

_____

_____

_____

_____

Mixed Review

Add or subtract.

11. $4 + 7 =$ _____

12. $14 - 7 =$ _____

13. $11 - 3 =$ _____

14. $7 + 9 =$ _____

15. $10 - 3 =$ _____

16. $9 + 5 =$ _____

17. $12 - 4 =$ _____

18. $5 + 6 =$ _____

19. $15 - 6 =$ _____

20. $17 - 8 =$ _____

21. $8 + 7 =$ _____

22. $6 + 4 =$ _____

Algebra: Missing Addends

Find the missing addend.

1. $9 + \underline{\hspace{1cm}} = 13$
2. $\underline{\hspace{1cm}} + 8 = 15$
3. $\underline{\hspace{1cm}} + 7 = 11$

4. $7 + \underline{\hspace{1cm}} = 12$
5. $9 + \underline{\hspace{1cm}} = 18$
6. $\underline{\hspace{1cm}} + 4 = 10$

7. $0 + \underline{\hspace{1cm}} = 7$
8. $\underline{\hspace{1cm}} + 8 = 14$
9. $7 + \underline{\hspace{1cm}} = 16$

10. $4 + \underline{\hspace{1cm}} = 12$
11. $8 + \underline{\hspace{1cm}} = 17$
12. $\underline{\hspace{1cm}} + 6 = 6$

13. $\underline{\hspace{1cm}} + 6 = 14$
14. $\underline{\hspace{1cm}} + 6 = 12$
15. $4 + \underline{\hspace{1cm}} = 13$

Write all the possible missing pairs of addends.

16. $\underline{\hspace{0.5cm}?\hspace{0.5cm}} + \underline{\hspace{0.5cm}?\hspace{0.5cm}} = 7$

_____

_____

_____

17. $\underline{\hspace{0.5cm}?\hspace{0.5cm}} + \underline{\hspace{0.5cm}?\hspace{0.5cm}} = 10$

_____

_____

_____

Mixed Review

Write the fact family for each set of numbers.

18. 7, 9, 16

19. 5, 6, 11

_____   _____

_____   _____

_____   _____

_____   _____

Name _____

Properties

Find each sum.

1. $9 + 4 =$ __13__   2. $(2 + 8) + 6 =$ __16__   3. $14 + 0 =$ __14__

    $4 + 9 =$ __13__ $2 + (8 + 6) =$ __16__

4. $7 + 8 =$ __15__   5. $0 + 13 =$ __13__   6. $5 + (4 + 7) =$ __16__

    $8 + 7 =$ __15__ $(5 + 4) + 7 =$ __16__

7. $4 + (8 + 5) =$ __17__   8. $18 + 0 =$ __18__   9. $9 + 6 =$ __15__

    $(4 + 8) + 5 =$ __17__ $6 + 9 =$ __15__

10. $0 + 15 =$ __15__   11. $5 + 7 =$ __12__   12. $(3 + 7) + 8 =$ __18__

                          $7 + 5 =$ __12__ $3 + (7 + 8) =$ __18__

13. $8 + 3 =$ __11__   14. $(1 + 5) + 7 =$ __13__   15. $9 + 0 =$ __9__

    $3 + 8 =$ __11__ $1 + (5 + 7) =$ __13__

16. $(2 + 3) + 4 =$ __9__   17. $0 + 17 =$ __17__   18. $9 + 1 =$ __10__

    $2 + (3 + 4) =$ __9__ $1 + 9 =$ __10__

Mixed Review

Add or subtract.

19. $16 - 9 =$ __7__   20. $8 + 5 =$ __13__   21. $7 + 5 =$ __12__

22. $11 - 5 =$ __6__   23. $17 - 8 =$ __9__   24. $5 + 9 =$ __14__

Find the missing addend.

25. $5 +$ __9__ $= 14$ 26. __8__ $+ 7 = 15$ 27. $7 +$ __7__ $= 14$

28. $9 +$ __3__ $= 12$ 29. __6__ $+ 5 = 11$ 30. __4__ $+ 9 = 13$

Name _____

Two-Digit Addition

Find the sum.

1. 54
    + 5
    ____
    59

2. 26
    + 73
    ____
    99

3. 18
    + 54
    ____
    72

4. 23
    38
    + 7
    ____
    68

5. 45
    + 42
    ____
    87

6. 37
    6
    + 84
    ____
    127

7. 79
    + 6
    ____
    85

8. 47
    + 89
    ____
    136

9. 36
    + 58
    ____
    94

10. 41
    + 9
    ____
    50

11. 83
    + 68
    ____
    151

12. 65
    39
    + 85
    ____
    189

13. 57
    + 42
    ____
    99

14. 63
    + 17
    ____
    80

15. 75
    + 46
    ____
    121

16. 55
    31
    + 26
    ____
    112

Complete each table.

	Add 20.	
17.	38	58
18.	6	26
19.	67	87

	Add 37.	
20.	46	83
21.	93	130
22.	77	114

	Add 52.	
23.	21	73
24.	65	117
25.	44	96

Mixed Review

Find the missing addend.

26. $6 + \underline{8} = 14$

27. $\underline{5} + 6 = 11$

28. $9 + \underline{8} = 17$

29. $4 + \underline{9} = 13$

30. $\underline{2} + 8 = 10$

31. $\underline{7} + 9 = 16$

32. $\underline{3} + 8 = 11$

33. $8 + \underline{8} = 16$

34. $6 + \underline{7} = 13$

Name _____

Two-Digit Subtraction

Find the difference. Use addition to check.

1.	48 − 5	**2.**	60 − 20	**3.**	59 − 46	**4.**	95 − 43

5.	35 − 9	**6.**	84 − 56	**7.**	87 − 8	**8.**	70 − 16

9.	86 − 28	**10.**	90 − 9	**11.**	62 − 26	**12.**	83 − 68

Complete each table.

	Subtract 40.	
13.	70	
14.	86	
15.	63	

	Subtract 28.	
16.	99	
17.	52	
18.	90	

Mixed Review

Write the fact family for each set of numbers.

19. 5, 8, 13   **20.** 8, 8, 16

_____   _____

_____   _____

_____   _____

_____   _____

Find the sum.

21.	35 + 63	**22.**	59 + 6	**23.**	78 + 27	**24.**	68 + 69

Problem Solving Skill

Choose the Operation

Choose the operation. Write a number sentence. Then solve.

1. Ashley is making a patchwork quilt. She cut out 46 squares. Then she cut out 25 triangles. How many quilt pieces did she cut out?

   _____

   _____

   _____

2. Van and Lisa were playing darts. Van scored 49 points, and Lisa scored 83 points. How many more points did Lisa score than Van?

   _____

   _____

   _____

For Exercises 3–4, use the following.

On Robin's farm there are 68 cows. In the spring 54 calves were born. How many cows and calves are there altogether?

3. What number sentence can you write to solve the problem?

   A $68 + 68 = 136$
   B $68 + 54 = 122$
   C $68 - 54 = 14$
   D $54 - 54 = 0$

4. How many more cows than calves are there?

   F 4
   G 12
   H 14
   J 16

Mixed Review

Solve.

5. $\begin{array}{r} 34 \\ + 13 \\ \hline \end{array}$

6. $\begin{array}{r} 86 \\ + 7 \\ \hline \end{array}$

7. $\begin{array}{r} 59 \\ - 36 \\ \hline \end{array}$

8. $\begin{array}{r} 86 \\ - 18 \\ \hline \end{array}$

9. $\begin{array}{r} 59 \\ + 41 \\ \hline \end{array}$

10. $\begin{array}{r} 43 \\ - 6 \\ \hline \end{array}$

11. $\begin{array}{r} 94 \\ - 38 \\ \hline \end{array}$

12. $\begin{array}{r} 74 \\ + 95 \\ \hline \end{array}$

Name _____

Even and Odd

1	2	3	4	5	6	7	8	9	10
11	12	13	14	15	16	17	18	19	20
21	22	23	24	25	26	27	28	29	30
31	32	33	34	35	36	37	38	39	40
41	42	43	44	45	46	47	48	49	50
51	52	53	54	55	56	57	58	59	60
61	62	63	64	65	66	67	68	69	70
71	72	73	74	75	76	77	78	79	80
81	82	83	84	85	86	87	88	89	90
91	92	93	94	95	96	97	98	99	100

Look at each number. Tell whether the number is *odd* or *even*.

1. 34   2. 15   3. 52   4. 23   5. 19

_____   _____   _____   _____   _____

6. 35   7. 82   8. 5   9. 89   10. 28

_____   _____   _____   _____   _____

Use the hundred chart.

11. Start at 2. Skip-count by twos. Move 12 skips. What number do you land on? Is it odd or even?

12. Start at 3. Skip-count by threes. Move 5 skips. What number do you land on? Is it odd or even?

_____   _____

Mixed Review

Find each sum or difference.

13.
$$\begin{array}{r} 63 \\ 45 \\ + 12 \\ \hline \end{array}$$

14.
$$\begin{array}{r} 35 \\ 43 \\ + 24 \\ \hline \end{array}$$

15.
$$\begin{array}{r} 26 \\ - 19 \\ \hline \end{array}$$

16.
$$\begin{array}{r} 86 \\ 41 \\ + 20 \\ \hline \end{array}$$

17. $18 + 22 + 19 =$ _____   18. $45 - 6 =$ _____   19. $32 + 18 =$ _____

© Harcourt

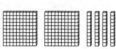

Place Value: 3-Digit Numbers

Write each number in standard form.

1. 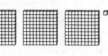 2.   3.

   _____   _____   _____

4. $300 + 40 + 9$ _____

5. $100 + 60 + 3$ _____

6. $700 + 90 + 9$ _____

7. seven hundred eighty _____

8. six hundred thirty-two _____

9. 5 hundreds 6 ones _____

10. two hundreds 4 tens eight ones _____

Write the value of the underlined digit.

11. 73<u>6</u> _____   12. <u>3</u>41 _____

13. 7<u>5</u>0 _____   14. <u>4</u>08 _____

Mixed Review

Add or subtract.

15.	82 -24	16.	34 $+56$	17.	35 $+\ 6$	18.	71 -42

19.	64 -28	20.	32 $-\ 7$	21.	18 $+18$	22.	88 $+15$

© Harcourt

Name _____

Algebra: Number Patterns

Predict the next number in each pattern.

1.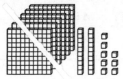

   _____ _____

2.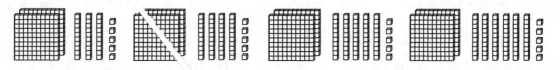

   _____

3. 120, 110, 100, 90, _____

4. 350, 400, 450, 500, _____

5. 680, 675, 670, 665, _____

6. 3,123; 4,123; 5,123; 6,123; _____

7. 409, 412, 415, 418, _____

8. 1,296; 1,294; 1,292; 1,290; _____

Mixed Review
Find the missing addend.

9. $6 + \underline{\hspace{1cm}} = 15$

10. $0 + \underline{\hspace{1cm}} = 8$

11. $\underline{\hspace{1cm}} + 9 = 16$

12. $\underline{\hspace{1cm}} + 7 = 7$

13. $\underline{\hspace{1cm}} + 8 = 13$

14. $8 + \underline{\hspace{1cm}} = 17$

Find each sum.

15. $8 + 6 = \underline{\hspace{1cm}}$

   $6 + 8 = \underline{\hspace{1cm}}$

16. $(6 + 4) + 3 = \underline{\hspace{1cm}}$

   $6 + (4 + 3) = \underline{\hspace{1cm}}$

17. $0 + 17 = \underline{\hspace{1cm}}$

18. $9 + 4 = \underline{\hspace{1cm}}$

   $4 + 9 = \underline{\hspace{1cm}}$

19. $15 + 0 = \underline{\hspace{1cm}}$

20. $6 + (2 + 8) = \underline{\hspace{1cm}}$

   $(6 + 2) + 8 = \underline{\hspace{1cm}}$

Place Value: 5- and 6-Digit Numbers

Write in standard form.

1. $30,000 + 5,000 + 300 + 20 + 1$

_____

2. $40,000 + 9,000 + 400 + 70 + 2$

_____

3. $700,000 + 20,000 + 3,000$

_____

4. $80,000 + 800 + 8$

_____

5. $70,000 + 200 + 80 + 9$

_____

6. $300,000 + 10,000 + 90 + 4$

_____

7. two hundred sixty-one
thousand, eight hundred
thirty-one

_____

8. forty-three thousand, five
hundred forty-five

_____

Write the value of the underlined digit.

9. 91,643

10. 536,955

11. 72,561

_____   _____   _____

12. 15,406

13. 21,789

14. 445,632

_____   _____   _____

Mixed Review

Solve.

15. $16 + 15 =$ _____

16. $20 - 17 =$ _____

17. $22 -$ _____ $= 14$

18. $17 + 58 =$ _____

19. $38 + 62 =$ _____

20. $40 - 36 =$ _____

21. $24 - 18 =$ _____

22. $16 + 16 =$ _____

Benchmark Numbers

Choose a benchmark of 10 or 100 to estimate each.

1. the number of doors in your home _____

2. the number of crackers in a large box _____

3. the number of hours in the school day _____

4. the number of pages in a book of sports stories _____

5. the number of players on a baseball team _____

Choose a benchmark of 25, 100, or 500 to estimate each.

6. the number of desks in your classroom _____

7. the number of seats in a high school sports stadium _____

8. the number of shopping carts at a large supermarket _____

9. the number of slices in a loaf of bread _____

10. the number of days in three months _____

Mixed Review

Add or subtract.

11.	71 −22	12.	95 +46	13.	82 −30	14.	39 +75
15.	66 +45	16.	26 +16	17.	50 −27	18.	45 +98
19.	79 −42	20.	88 −65	21.	80 +44	22.	92 −75

Algebra: Compare Numbers

Compare the numbers. Write $<$, $>$, or $=$ in the ◯.

1. 256 ◯ 266

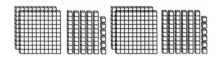

2. 138 ◯ 136

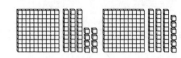

3. 1,231 ◯ 1,123

4. 2,045 ◯ 2,055

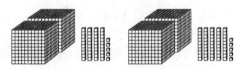

5. 85,604 ◯ 85,604

6. 44,444 ◯ 444,444

7. 36,542 ◯ 36,245

8. 814,365 ◯ 804,365

Mixed Review

Write the number in standard form.

9. 40,000 + 6,000 + 300 + 50 + 5 _____

10. 200,000 + 7,000 + 600 + 20 + 9 _____

11. eight thousand, three hundred fifty-two _____

12. forty-three thousand, six hundred twenty-five _____

Write the number in expanded form.

13. 17,045 _____

14. 596,811 _____

15. 4,906 _____

Complete the pattern.

16. 25, 30, 35, ____, ____

17. 17, 20, 23, ____, ____

18. 152, 252, 352, ____, ____

19. 79, 69, 59, ____, ____

Order Numbers

Write the numbers in order from *least* to *greatest*.

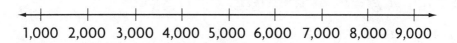

1,000 2,000 3,000 4,000 5,000 6,000 7,000 8,000 9,000

1. 2,221; 2,210; 2,235   **2.** 4,305; 3,275; 3,255   **3.** 7,246; 7,232; 7,310

_____   _____   _____

4. 2,326; 1,503; 3,235   **5.** 5,609; 5,950; 4,999   **6.** 9,000; 7,607; 4,439

_____   _____   _____

7. 8,256; 6,208; 7,065   **8.** 4,135; 2,857; 4,351   **9.** 2,904; 2,499; 1,894

_____   _____   _____

Write the numbers in order from *greatest* to *least*.

10. 1,652; 1,328; 1,691   **11.** 87,114; 88,205; 79,343

_____   _____

12. 254,357; 124,899; 304,506

_____

Mixed Review

Solve.

13. $29 + 10 + 4 =$ _____   **14.** $71 + 12 + 8 =$ _____

15. $53 + 11 + 14 =$ _____   **16.** $72 + 8 + 0 =$ _____

17. $13 + 58 + 29 =$ _____   **18.** $49 + 49 + 10 =$ _____

19. $\begin{array}{r} 71 \\ -39 \\ \hline \end{array}$ **20.** $\begin{array}{r} 97 \\ -38 \\ \hline \end{array}$ **21.** $\begin{array}{r} 91 \\ -84 \\ \hline \end{array}$ **22.** $\begin{array}{r} 60 \\ -25 \\ \hline \end{array}$

Problem Solving Skill

Use a Bar Graph

For 1–2, use the bar graph at the right.

1. Peggy's popcorn machine can make about 10,000 bags of popcorn a week. For which types of popcorn would it take more than a week to make all the bags?

   _____

2. One tub of kernels can make about 1,000 bags of popcorn. How many tubs of kernels does Peggy need to make caramel popcorn? Explain.

   _____

   _____

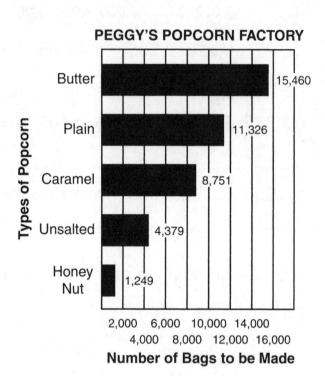

PEGGY'S POPCORN FACTORY

Types of Popcorn

Butter — 15,460
Plain — 11,326
Caramel — 8,751
Unsalted — 4,379
Honey Nut — 1,249

2,000   6,000   10,000   14,000
  4,000   8,000   12,000   16,000

Number of Bags to be Made

Mixed Review

Write <, >, or = in the ◯.

3. 3,456 ◯ 346

4. 1,216 ◯ 1,154

5. 7,756 ◯ 7,776

6. 84,448 ◯ 84,448

7. 19,213 ◯ 91,213

8. 365,251 ◯ 365,215

Solve.

9.   34
   − 15

10.   47
    + 14

11.   78
    − 39

12.   26
    + 34

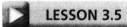

Round to Nearest 10 and 100

Round to the nearest ten.

1. 26 _____   **2.** 85 _____   **3.** 72 _____   **4.** 55 _____

5. 17 _____   **6.** 31 _____   **7.** 88 _____   **8.** 97 _____

9. 46 _____   **10.** 62 _____   **11.** 8 _____   **12.** 29 _____

Round to the nearest hundred and the nearest ten.

13. 564 _____ _____   **14.** 412 _____ _____

15. 625 _____ _____   **16.** 445 _____ _____

17. 454 _____ _____   **18.** 621 _____ _____

19. 533 _____ _____   **20.** 689 _____ _____

21. 599 _____ _____   **22.** 327 _____ _____

23. 555 _____ _____   **24.** 649 _____ _____

Mixed Review

Tell whether the number is *odd* or *even*.

25. 1,784 _____   **26.** 333 _____   **27.** 95 _____

28. 178 _____   **29.** 712 _____   **30.** 619 _____

Solve.

31. $90 - 12 =$ _____   **32.** $39 + 21 =$ _____

33. $47 + 54 =$ _____   **34.** $60 - 23 =$ _____

35. $\begin{array}{r} 93 \\ -\ 78 \\ \hline \end{array}$ **36.** $\begin{array}{r} 56 \\ -\ 48 \\ \hline \end{array}$ **37.** $\begin{array}{r} 57 \\ +\ 63 \\ \hline \end{array}$ **38.** $\begin{array}{r} 82 \\ -\ 39 \\ \hline \end{array}$

Round to Nearest 1,000

Round to the nearest thousand.

1. 2,345 _____   **2.** 1,765 _____   **3.** 8,821 _____

4. 6,109 _____   **5.** 3,001 _____   **6.** 3,679 _____

7. 9,134 _____   **8.** 4,556 _____   **9.** 7,733 _____

Round to the nearest thousand, the nearest hundred, and the nearest ten.

10. 3,490 _____ _____ _____

11. 7,509 _____ _____ _____

12. 2,565 _____ _____ _____

13. 3,115 _____ _____ _____

14. 1,350 _____ _____ _____

15. 8,999 _____ _____ _____

16. 6,784 _____ _____ _____

17. 2,288 _____ _____ _____

18. 5,501 _____ _____ _____

Mixed Review

Write the value of the underlined digit.

19. 4,5_2_3 _____   **20.** _1_3,886 _____   **21.** 60,_6_00 _____

22. _3_27 _____   **23.** 68_7_,235 _____   **24.** _2_2,789 _____

Solve.

25.	65	**26.**	86	**27.**	49	**28.**	92
	−48		−58		+13		−34

Name _____

Estimate Sums

Use rounding to estimate the sum.

1.	236	2.	$4.84	3.	6,927	4.	$42.98
	+ 710		+ $2.63		+ 1,280		+ $25.79

Use front-end estimation to estimate the sum.

5.	436	6.	$1.82	7.	3,467	8.	$12.52
	+ 517		+ $2.64		+ 7,517		+ $28.34

For 9–11, use the numbers at the right.

9. Choose two numbers whose sum is about 80.

_____

10. Choose two numbers whose sum is about 4,000.

_____

11. Choose two numbers whose sum is about 700.

_____

| 533 |
| 38 |
| 1,092 |
| 41 |
| 229 |
| 3,481 |

Mixed Review

Write $<$, $>$, or $=$ for each $\bigcirc$.

12. 334 $\bigcirc$ 443   13. 4,980 $\bigcirc$ 4,098

14. 814 $\bigcirc$ 814   15. 39,215 $\bigcirc$ 31,872

Write each number in standard form.

16. $60,000 + 2,000 + 500 + 50$ _____

17. forty-three thousand, nine hundred sixty-six _____

18. $2,000 + 900 + 40 + 3$ _____

19. eight hundred thousand, two hundred eleven _____

20. $700,000 + 3,000 + 200 + 70 + 9 =$ _____

Addition with Regrouping

Find each sum.

1. 341
  +237

2. 832
  +138

3. 426
  +427

4. 359
  +196

5. 532
  +389

6. 644
  +317

7. 277
  +235

8. 442
  +469

9. 353
  +588

10. 527
   +197

11. 438
   +279

12. 377
   +195

13. 159
   +262

14. 349
   +464

15. 618
   +329

16. 627
   +326

17. 378
   +577

18. 819
   +153

19. 377
   +188

20. 429
   +469

Mixed Review

Add.

21. 57
   +36

22. 88
   +97

23. 49
   +57

24. 67
   +38

25. 49
   +89

Subtract.

26. 71
   −32

27. 98
   −84

28. 83
   −57

29. 56
   −38

30. 99
   −81

31. 92
   −18

32. 14
   − 8

33. 76
   −54

34. 29
   −14

35. 75
   −26

© Harcourt

Add 3- and 4-Digit Numbers

Find the sum. Estimate to check.

1. 356
 +228

2. $14.95
 +$22.78

3. 657
 +155

4. 1,494
 +9,369

5. 4,364
 +2,465

6. 7,648
 +5,173

7. $64.93
 +$34.82

8. 146
 +594

9. $52.47
 +$34.53

10. 152
 +688

11. $38.46
 +$16.59

12. 473
 +437

13. 3,349
 +8,449

14. 147
 +366

15. 528
 869
 +131

Mixed Review

Write the value of the underlined digit.

16. 2<u>5</u>,781

17. <u>1</u>3,499

18. <u>2</u>45,006

19. <u>7</u>7,712

_____ _____ _____ _____

20. <u>5</u>76

21. 92,4<u>4</u>0

22. 11,29<u>9</u>

23. 4,<u>8</u>10

_____ _____ _____ _____

Round to the nearest ten.

24. 566

25. 717

26. 32

27. 673

_____ _____ _____ _____

28. 1,854

29. 392

30. 428

31. 4,668

_____ _____ _____ _____

Problem-Solving Strategy

Predict and Test

Use *predict and test* to solve.

1. Two numbers have a sum of 39. Their difference is 11. What are the two numbers?

   _____

2. Two numbers have a sum of 22. Their difference is 4. What are the two numbers?

   _____

3. Gina traveled 450 miles to her grandmother's house in two days. She traveled 50 more miles on Saturday than on Sunday. How many miles did she travel on Saturday? on Sunday?

   _____

4. Maria practiced the recorder for 40 minutes on Saturday. She practiced 10 minutes less in the afternoon than in the morning. How many minutes did Maria practice in the morning? in the afternoon?

   _____

Mixed Review

Solve.

5. $17 + 22 + 56 =$ _____

6. $\$42.80 + \$23.90 + \$6.00 =$ _____

7. $134 + 326 + 422 =$ _____

8. $79 + 18 + 27 =$ _____

Write $<$, $>$, or $=$ in the $\bigcirc$.

9. $25 + 25 \bigcirc 50$

10. $721 + 322 \bigcirc 1{,}000$

11. $\$3.50 + \$2.25 \bigcirc \$4.25$

12. $582 + 241 \bigcirc 1{,}200$

13. $276 + 524 \bigcirc 800$

14. $\$19.83 + \$4.99 \bigcirc \$25.00$

Solve.

15. $\begin{array}{r} 19 \\ +59 \\ \hline \end{array}$

16. $\begin{array}{r} 276 \\ +347 \\ \hline \end{array}$

17. $\begin{array}{r} 365 \\ +485 \\ \hline \end{array}$

18. $\begin{array}{r} 63 \\ 29 \\ +15 \\ \hline \end{array}$

19. $\begin{array}{r} 54 \\ 48 \\ +39 \\ \hline \end{array}$

Name _____

Choose a Method

Find the sum. Tell what method you used.

1.   2,341
  +6,237
	2. 861
	  +733
		3. 800
		  +300
			4. 1,776
			  +1,954

5.   1,952
  +1,980
	6. 988
	  +982
		7. 1,113
		  +5,988
			8. $7.82
			  +$9.39

9.   4,000
  +3,000
	10. 6,318
	  +4,916
		11. 7,657
		  +1,284
			12. 5,000
			  +8,000

13.   588
  +455
	14. 5,387
	  +8,347
		15. $4.25
		  +$5.56
			16. 6,859
			  +1,346

Mixed Review

Write the numbers in order from *least* to *greatest*.

17. 245, 253, 232   18. 7,924; 7,429; 7,249   19. 632, 599, 900

_____   _____   _____

Add.

20.   47
     69
  +81
	21. 75
	     83
	  +52
		22. 94
		     18
		  +60
			23. 26
			     99
			  +34

24.   221
  +876
	25. 595
	  +111
		26. 469
		  +568
			27. 670
			  +710

Algebra: Expressions and Number Sentences

Write an expression. Then write a number sentence to solve.

1. Garnet bought 16 red buttons, 8 blue buttons, and 25 green buttons. How many blue and red buttons did she buy?

_____

2. Kay has 13 more sheets of lined paper than unlined paper. She has 26 sheets of unlined paper. How many sheets of lined paper does she have?

_____

3. Lyle had 152 pages to read in his library book. He read 65 pages. How many pages does he have left?

_____

4. Neil had 35 cookies. He gave 26 cookies to his classmates. How many cookies does he have left?

_____

Write + or − to complete the number sentence.

5. $4 \bigcirc 2 = 2$

6. $27 = 18 \bigcirc 9$

7. $32 \bigcirc 3 = 35$

8. $67 = 7 \bigcirc 60$

9. $39 \bigcirc 16 = 55$

10. $16 \bigcirc 11 = 5$

11. $15 \bigcirc 7 = 8$

12. $50 = 61 \bigcirc 11$

13. $71 = 43 \bigcirc 28$

Write the missing number.

14. $9 + \underline{\hphantom{xx}} = 21$

15. $8 = \underline{\hphantom{xx}} - 9$

16. $\underline{\hphantom{xx}} + 81 = 93$

17. $160 = 50 + \underline{\hphantom{xx}}$

18. $\underline{\hphantom{xx}} - 123 = 16$

19. $36 - \underline{\hphantom{xx}} = 5$

20. $57 + 18 = \underline{\hphantom{xx}}$

21. $115 - 113 = \underline{\hphantom{xx}}$

22. $237 - \underline{\hphantom{xx}} = 195$

Mixed Review

Find each sum.

23.
$$\begin{array}{r} 25 \\ 70 \\ +97 \\ \hline \end{array}$$

24.
$$\begin{array}{r} 38 \\ 63 \\ +81 \\ \hline \end{array}$$

25.
$$\begin{array}{r} 52 \\ 49 \\ +74 \\ \hline \end{array}$$

26.
$$\begin{array}{r} 86 \\ 85 \\ +38 \\ \hline \end{array}$$

Name _____

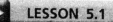

Estimate Differences

Use rounding to estimate the difference.

1. $59 \rightarrow$ _____
    $-\ 16 \rightarrow -$ _____

2. $\$8.17 \rightarrow$ _____
    $-\ \$5.51 \rightarrow -$ _____

3. $8{,}909 \rightarrow$ _____
    $-\ 2{,}408 \rightarrow -$ _____

Use front-end estimation to estimate the difference.

4. $83 \rightarrow$ _____
    $-\ 38 \rightarrow -$ _____

5. $5{,}501 \rightarrow$ _____
    $-\ 3{,}288 \rightarrow -$ _____

6. $\$8.15 \rightarrow$ _____
    $-\ \$4.37 \rightarrow -$ _____

Estimate the difference. Round to the place that makes sense.

7. $728 \rightarrow$ _____
    $-\ 684 \rightarrow -$ _____

8. $504 \rightarrow$ _____
    $-\ 467 \rightarrow -$ _____

9. $8{,}316 \rightarrow$ _____
    $-\ 7{,}923 \rightarrow -$ _____

Mixed Review

Write the missing number.

10. 8, 13, _____, 23, 28

11. 16, 23, 30, 37, _____

12. _____, 20, 29, 38, 47

Write the value of the underlined digit.

13. 5̲3,980 _____

14. 46,8̲31 _____

15. $3̲67.15 _____

Add.

16. $3{,}483$
     $+\ \ \ 547$

17. $1{,}209$
     $+\ \ \ 593$

18. $1{,}756$
     $+\ 8{,}394$

19. $7{,}674$
     $+\ 3{,}421$

20. $54 + 24 + 17 =$ _____

21. $31 + 31 + 39 =$ _____

22. $35 + 26 + 13 =$ _____

23. $42 + 63 + 12 =$ _____

Subtraction with Regrouping

Use base-ten blocks. Draw a picture to show the difference.

1. $352 - 236 =$ _____   **2.** $532 - 248 =$ _____   **3.** $436 - 127 =$ _____

4. $457 - 285 =$ _____   **5.** $512 - 369 =$ _____   **6.** $327 - 127 =$ _____

7. $438 - 249 =$ _____   **8.** $367 - 175 =$ _____   **9.** $452 - 259 =$ _____

10. $414 - 126 =$ _____   **11.** $378 - 187 =$ _____   **12.** $333 - 155 =$ _____

Mixed Review

Add.

13.	14.	15.	16.	17.
150 $+\ 30$	60 $+90$	72 $+35$	56 $+28$	165 $+\ 67$

Subtract.

18.	19.	20.	21.	22.
80 -30	90 -50	79 -24	84 -57	91 -37

23.	24.	25.	26.	27.
73 -35	65 -16	34 -17	62 -28	71 -14

© Harcourt

Subtract Across Zeros

Find the difference. Estimate to check.

1. 354 −148	2. 500 −139	3. 942 −817	4. $7.83 −$5.26	5. 604 −435

6. 305 −178	7. 635 −145	8. 700 − 58	9. $8.00 −$4.59	10. 461 −178

11. 401 −275	12. $9.21 −$7.32	13. 437 −128	14. 675 −179	15. 700 −536

16. 729 −518	17. $4.06 −$2.97	18. $9.00 −$6.95	19. 500 − 26	20. 372 −158

Mixed Review

21. 119 +669	22. 542 +669	23. 908 +103	24. 275 +479

25. 77 −12	26. 48 −15	27. 95 −37	28. 41 − 8

29. 603 +279	30. 400 +118	31. 525 +175	32. 235 + 66

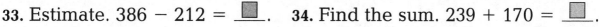

33. Estimate. $386 − 212 =$ ▨.

    A 100   **C** 300

    B 200   **D** 500

34. Find the sum. $239 + 170 =$ ▨.

    F 400   **H** 409

    G 308   **J** 309

© Harcourt

Name _____

Subtract 3- and 4-Digit Numbers

Find the difference. Estimate to check.

1.	2.	3.	4.	5.
624 −471	3,917 −1,653	$50.84 −$28.37	892 −395	7,301 −2,285

6.	7.	8.	9.	10.
2,563 − 886	$41.95 −$25.77	8,007 −3,456	3,916 −1,867	$52.42 − $ 7.99

11.	12.	13.	14.	15.
9,050 −4,283	4,328 −3,759	$96.19 −$27.50	7,000 −5,432	3,284 −1,586

Write <, >, or =.

16. 2,072 − 1,947 ◯ 703 − 568   17. 7,107 − 4,726 ◯ 8,264 − 5,883

18. 831 − 269 ◯ 2,307 − 1,845   19. 6,267 − 2,489 ◯ 1,407 − 929

Mixed Review

Write the numbers in order from *least* to *greatest*.

20. 325, 315, 345   21. 4,683; 3,527; 2,904

_____   _____

22. 76,218; 86,374; 79,506   23. 35,241; 35,412; 35,142

_____   _____

Write the numbers in order from *greatest* to *least*.

24. 3,729; 2,846; 3,851   25. 56,084; 61,326; 66,100

_____   _____

26. 472,163; 539,208; 341,957   27. 451,925; 471,922; 417,922

_____   _____

Choose a Method

Find the difference. Tell what method you used.

1. 1,500	2. 406	3. 1,600	4. 2,902	5. 700
−1,132	−258	−1,198	−2,435	−137

6. 3,408	7. 800	8. 3,306	9. 6,300	10. 8,000
−2,135	−600	−3,108	−2,229	−5,000

11. 7,005 − 3,605 = _____   **12.** 8,588 − 5,666 = _____

13. 2,175 − 1,987 = _____   **14.** 921 − 108 = _____

Mixed Review

Find each sum or difference.

15. 19 + 6 = _____   **16.** 78 − 49 = _____

17. 84 − 27 = _____   **18.** 29 + 54 = _____

Find the missing addend.

19. 24 + _____ = 60   **20.** 35 + _____ = 71

21. 17 + _____ = 58   **22.** 42 + _____ = 79

Find each sum.

23. 996 + 132 = _____   **24.** 4,597 + 1,950 = _____

25. 3,956 + 2,007 = _____   **26.** 774 + 2,981 = _____

27. Which number is between 4,888 and 6,123?

    A 5,030   **C** 1,325

    B 7,548   **D** 3,987

28. Which symbol completes the following:

    4,620 ◯ 4,062

    F >   **G** <   **H** =

Problem Solving Skill

Estimate or Exact Answer

Use the table for 1–2. Write whether you need an exact answer or an estimate. Then solve.

Camping Supplies	
Item	**Price**
Cooler	$36.29
Lantern	$23.88
Sleeping bag	$74.99

1. Justin has $100. Can he buy a cooler and a sleeping bag? Explain.

   _____

   _____

2. Roxana pays for a lantern with $30. How much change will she get?

   _____

There will be 258 adults and 362 children at the Lazy River Campground this weekend. The campground will give one trail map to each camper. How many maps are needed in all?

3. Which number sentence can you use to solve the problem?

   A $258 + 362 =$ _____
   B $300 + 400 =$ _____
   C $300 + 362 =$ _____
   D $362 - 258 =$ _____

4. How many maps does the campground need in all?

   F 104 maps
   G 610 maps
   H 620 maps
   J 700 maps

Mixed Review

Solve.

5. $\begin{array}{r} 3,641 \\ -2,915 \\ \hline \end{array}$

6. $\begin{array}{r} 1,094 \\ +6,378 \\ \hline \end{array}$

7. $\begin{array}{r} 5,183 \\ -4,692 \\ \hline \end{array}$

8. $\begin{array}{r} 2,796 \\ +5,847 \\ \hline \end{array}$

Name _____

Compare Money Amounts

Use < or > to compare the amounts of money.

1. a.    **b.**

_____

2. a.    **b.**

_____

3. a.      **b.**

_____

Mixed Review

4. Continue the pattern.

19, 29, 39, 49, _____, _____, _____

Find the sum.

5.	6.	7.	8.
85	14	565	26
72	33	+ 128	38
+ 21	+ 67		+ 52

9. What is the value of the underlined digit in 10,7̲29?

A 70   **C** 7,000

B 700   **D** 70,000

10. What is the value of the underlined digit in 1̲8,246?

A 80   **C** 8,000

B 800   **D** 80,000

© Harcourt

Name _____

Make Change

List the coins you would get as change from a $5 bill.
Use play money.

1. $4.92

2. $3.35

3. $2.59

_____ _____ _____

_____ _____ _____

Complete the table. Use play money.

	COST OF ITEM	AMOUNT PAID	CHANGE IN BILLS AND COINS	TOTAL AMOUNT OF CHANGE
4.	$0.19	$1.00	_____ _____ _____	_____
5.	$2.73	$5.00	_____ _____ _____	_____
6.	$5.31	$10.00	_____ _____ _____	_____

Mixed Review

Find the sum or difference.

7. 264
  + 599

8. 3,672
  − 1,488

9. 4,628
  − 1,999

10. 2,870
  + 9,653

11. Order these numbers from least to greatest.

3,876   3,678   3,768

_____

12. What is one hundred more than 7,409?

_____

© Harcourt

Add and Subtract Money

Find the sum or difference. Estimate to check.

1.	$6.43 +$2.15	2.	$5.63 −$1.50	3.	$2.59 +$1.37	4.	$4.93 −$1.78

5.	$0.38 +$5.24	6.	$3.27 +$2.06	7.	$6.55 −$4.90	8.	$4.02 −$3.91

9.	$3.50 −$1.98	10.	$1.90 +$2.64	11.	$63.94 +$32.78	12.	$28.06 +$52.44

13.	$19.78 +$53.98	14.	$50.00 −$19.89	15.	$75.45 −$36.47	16.	$82.02 −$75.93

Mixed Review

Write the missing number.

17. _____ tens = 50

18. _____ hundreds = 300

19. _____ tens = 90

20. _____ thousands = 6,000

21. _____ dimes = 4 quarters

22. 15 pennies = _____ dimes

_____ pennies

23. 12 dimes = _____ dollars

_____ dimes

24. 8 dimes = _____ quarters

_____ dimes

25. 26 nickels = _____ dollars

_____ dimes

26. 15 dimes = _____ dollars

_____ quarters

© Harcourt

Tell Time

Write each time. Then write two ways you can read each time.

1.

_____

_____

_____

2.

_____

_____

_____

Write two ways you can read each time.

3.

`9:17`

_____

_____

4.

`3:31`

_____

_____

Estimate each time to the nearest half hour.

5.

_____

6.

_____

7.

_____

Mixed Review

8.
```
  632
  421
+ 267
```

9.
```
  2,345
  1,827
+ 4,558
```

10.
```
  4,414
− 3,399
```

11.
```
  7,212
− 3,946
```

Name _____

A.M. and P.M.

Write the time, using A.M. or P.M.

1.

still sleeping

_____

2.
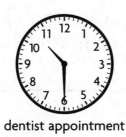
dentist appointment

_____

3.

paint a picture

_____

4.

lunch time

_____

5.

the sunrise

_____

6.

this is a new day

_____

7.

this day is almost over

_____

8.

do the dishes

_____

9.

eat breakfast

_____

Mixed Review

Write + or − to make the number sentence true.

10. $36 \bigcirc 27 = 9$

11. $16 = 14 \bigcirc 2$

12. $35 \bigcirc 18 = 53$

13. $15 = 22 \bigcirc 7$

Subtract.

14. $1.68
   −$0.09

15. $5.62
   −$3.17

16. $8.13
   −$3.59

17. $12.72
   −$ 7.49

Elapsed Time

Use a clock to find the elapsed time.

1. start: 4:15 P.M.
end: 4:30 P.M.

2. start: 5:30 P.M.
end: 7:50 P.M.

3. start: 3:30 A.M.
end: 4:15 A.M.

_____   _____   _____

Use a clock to find the end time.

4. starting time: 4:15 P.M.
elapsed time: 30 minutes

5. starting time: 2:00 A.M.
elapsed time: 3 hours and
30 minutes

_____   _____

6. starting time: 7:30 A.M.
elapsed time: 45 minutes

7. starting time: 3:45 P.M.
elapsed time: 5 hours

_____   _____

Mixed Review

Write $<$, $>$, or $=$ in each $\bigcirc$.

8. 1,980 $\bigcirc$ 1,980

9. 13,886 $\bigcirc$ 13,688

10. 6,807 $\bigcirc$ 6,870

11. 499 $-$ 107 $\bigcirc$ 307

Write in standard form.

12. six thousand, three hundred forty-two _____

13. 10,000 + 5,000 + 900 + 30 + 2 _____

14. 20,000 + 7,000 + 400 + 80 + 7 _____

15. eighty-four thousand, thirty-three _____

© Harcourt

Problem Solving Skill

Sequence Events

Use the table at the right.

1. Add points to the time line for each event listed in the table.

EVENTS IN SPACE

EVENT	YEAR
Hubble telescope launched	1990
Discovery first launched	1984
First space station	1973
First shuttle flight	1981
First American in space	1961
First moon walk	1969

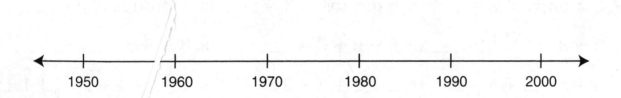

1950   1960   1970   1980   1990   2000

For 2–4, use your time line.

2. In 1983 Sally Ride became the first American woman to go into space. Between which two events should this event be on your time line?

_____

3. Which event occurred before the first space station?

   A Hubble telescope launched
   B First moon walk
   C First shuttle flight
   D *Discovery* first launched

4. Which event occurred after the first shuttle flight?

   F First space station
   G First American in space
   H First moon walk
   J Hubble telescope launched

Mixed Review

5. $1,750 + $ _____ $ = 6,750$ 6. $608 - $ _____ $ = 352$ 7. $2,374 + $ _____ $ = 9,815$

Algebra: Connect Addition and Multiplication

For 1–4, choose the letter of the number sentence that matches.

1. $6 + 6 + 6 + 6 + 6 = 30$ _____

2. $4 + 4 + 4 + 4 + 4 + 4 + 4 + 4 = 32$ _____

3. $5 + 5 + 5 + 5 = 20$ _____

4. $2 + 2 + 2 + 2 + 2 + 2 + 2 + 2 + 2 + 2 = 20$ _____

A $8 \times 4 = 32$

B $10 \times 2 = 20$

C $5 \times 6 = 30$

D $4 \times 5 = 20$

For 5–22, find the total. You may wish to draw a picture.

5. 2 groups of 6 = ___   6. 3 groups of 5 = ___   7. 2 groups of 4 = ___

8. 5 groups of 2 = ___   9. 6 groups of 3 = ___   10. 7 groups of 3 = ___

11. $3 + 3 + 3 + 3 =$ ___   12. $6 + 6 + 6 =$ ___   13. $8 + 8 =$ ___

14. $5 + 5 + 5 + 5 + 5$ 15. $2 + 2 + 2 + 2$ 16. $1 + 1 + 1 + 1 + 1 + 1$

= ___   = ___   = ___

17. $6 \times 1 =$ ___   18. $3 \times 2 =$ ___   19. $2 \times 9 =$ ___

20. $7 \times 2 =$ ___   21. $1 \times 7 =$ ___   22. $5 \times 5 =$ ___

Mixed Review

Write the missing number that makes the sentence true.

23. $4 +$ ___ $= 16$ 24. $5 =$ ___ $- 3$ 25. ___ $+ 16 = 22$ 26. $130 = 100 +$ ___

27. ___ $+ 7 = 23$ 28. $12 +$ ___ $= 30$ 29. $15 =$ ___ $+ 2$ 30. $70 +$ ___ $= 85$

Add.

31. $\begin{array}{r} 28 \\ + 17 \\ \hline \end{array}$ 32. $\begin{array}{r} 156 \\ + 813 \\ \hline \end{array}$ 33. $\begin{array}{r} 1{,}608 \\ + 1{,}097 \\ \hline \end{array}$ 34. $\begin{array}{r} 3{,}499 \\ + 3{,}499 \\ \hline \end{array}$

35. $\begin{array}{r} 362 \\ + 412 \\ \hline \end{array}$ 36. $\begin{array}{r} 2{,}130 \\ + 9{,}805 \\ \hline \end{array}$ 37. $\begin{array}{r} 4{,}091 \\ + 1{,}904 \\ \hline \end{array}$ 38. $\begin{array}{r} 2{,}694 \\ + 1{,}739 \\ \hline \end{array}$

Multiply with 3

Use the number line to find the product.

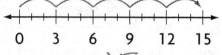

1. $5 \times 3 =$ 15

2. $3 \times 5 =$ 15

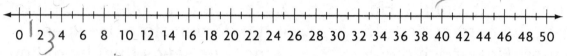

3. $5 \times 5 =$ 25   4. $4 \times 3 =$ ___   5. $9 \times 3 =$ ___   6. $2 \times 3 =$ ___

7. $4 \times 5 =$ ___   8. $3 \times 8 =$ ___   9. $7 \times 2 =$ ___   10. $3 \times 3 =$ ___

11. $9 \times 5 =$ ___   12. $6 \times 3 =$ ___   13. $2 \times 2 =$ ___   14. $5 \times 3 =$ ___

15. $8 \times 2 =$ ___   16. $5 \times 9 =$ ___   17. $2 \times 9 =$ ___   18. $6 \times 5 =$ ___

19. $5 \times 4 =$ ___   20. $3 \times 9 =$ ___   21. $5 \times 2 =$ ___   22. $7 \times 3 =$ ___

23. $8 \times 5 =$ ___   24. $7 \times 5 =$ ___   25. $2 \times 5 =$ ___

26. $5 \times 8 =$ ___   27. $3 \times 4 =$ ___   28. $2 \times 7 =$ ___

29. $3 \times 6 =$ ___   30. $9 \times 2 =$ ___   31. $8 \times 4 =$ ___

Mixed Review

Circle the letter for the correct answer.

32. $24 + 56 + 12 =$ ■
   A 29   **C** 101
   B 82   **D** 92

33. $17 + 11 + 45 =$ ■
   F 53   **H** 84
   G 73   **J** 102

34. $12 + 9 + 19 =$ ■
   A 40   **C** 45
   B 42   **D** 49

35. $62 + 15 + 27 =$ ■
   F 88   **H** 104
   G 92   **J** 114

36. $25 + 35 + 45 =$ ■
   A 75   **C** 90
   B 85   **D** 105

37. $26 + 38 + 7 =$ ■
   F 69   **H** 78
   G 71   **J** 81

Problem Solving Skill

Too Much/Too Little Information

Garden Supplies	
hoe	$9
rake	$8
package of seeds	$2

For 1–6, use the table.

For 1–4, write *a, b,* or *c* to tell whether the problem has *a.* too much information, *b.* too little information, or *c.* the right amount of information. Solve those with too much or the right amount of information.

1. Mario bought 2 rakes. He was in the garden store 15 minutes. How much did Mario spend?

_____a_____

2. Cecil left at 5:00 P.M. to go to the garden store. He spent more on seeds than he did on other garden supplies. How much did he spend on seeds?

_____

3. Jerome had $20. He bought 7 packages of seeds. How much did he spend?

_____

4. Elaine had $20. She bought one hoe and two shovels. How much did she spend?

_____

5. You have $25 to spend on garden supplies. Which items can you buy?

   A 2 hoes, 2 rakes

   B 3 rakes, a package of seeds

   C 2 hoes, 4 packages of seeds

   D 1 hoe, 2 rakes

6. You have $30. How much more money do you need if you choose to buy 4 packages of seeds, 2 rakes and 2 hoes?

   F $42   H $12

   G $13   J $10

Mixed Review

Write the time.

7.

8.

9.

10.

11. Are the hours between midnight and noon A.M. or P.M.? _____

Multiply with 1 and 0

Complete the multiplication sentence to show the number of sneakers.

 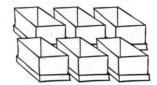

1. $3 \times 1 =$ _____   **2.** $6 \times 0 =$ _____   **3.** $1 \times 2 =$ _____

Find the product.

4. $8 \times 0 =$ _____   **5.** $1 \times 6 =$ _____   **6.** $0 \times 5 =$ _____   **7.** $9 \times 1 =$ _____

8. $1 \times 4 =$ _____   **9.** $0 \times 3 =$ _____   **10.** $1 \times 8 =$ _____   **11.** $0 \times 1 =$ _____

12. $0 \times 0 =$ _____   **13.** $5 \times 1 =$ _____   **14.** $7 \times 0 =$ _____   **15.** $2 \times 5 =$ _____

16. $5 \times 4 =$ _____   **17.** $6 \times 3 =$ _____   **18.** $3 \times 7 =$ _____   **19.** $8 \times 2 =$ _____

Mixed Review

20. Find the value of the bold digit.

43,9**7**5 _____   7**8**,214 _____

90,255 _____   33,4**3**6 _____

29,4**6**7 _____   89,**6**12 _____

21. Find the sum of 198 and 864. _____

22. Put the numbers in order from least to greatest.

    74   44   62   47

_____

23. Put the numbers in order from greatest to least.

    29   59   13   68

_____

24. $3 + 3 + 3 + 3 =$ _____   **25.** $2 + 2 + 2 =$ _____

Multiply with 4 on a Multiplication Table

Find the product.

1. $\begin{array}{r} 4 \\ \times\,4 \\ \hline \end{array}$
2. $\begin{array}{r} 1 \\ \times\,4 \\ \hline \end{array}$
3. $\begin{array}{r} 4 \\ \times\,7 \\ \hline \end{array}$
4. $\begin{array}{r} 9 \\ \times\,4 \\ \hline \end{array}$
5. $\begin{array}{r} 4 \\ \times\,3 \\ \hline \end{array}$
6. $\begin{array}{r} 2 \\ \times\,4 \\ \hline \end{array}$
7. $\begin{array}{r} 4 \\ \times\,8 \\ \hline \end{array}$

8. $\begin{array}{r} 0 \\ \times\,4 \\ \hline \end{array}$
9. $\begin{array}{r} 5 \\ \times\,4 \\ \hline \end{array}$
10. $\begin{array}{r} 3 \\ \times\,2 \\ \hline \end{array}$
11. $\begin{array}{r} 4 \\ \times\,2 \\ \hline \end{array}$
12. $\begin{array}{r} 4 \\ \times\,1 \\ \hline \end{array}$
13. $\begin{array}{r} 7 \\ \times\,3 \\ \hline \end{array}$
14. $\begin{array}{r} 9 \\ \times\,2 \\ \hline \end{array}$

15. $\begin{array}{r} 8 \\ \times\,2 \\ \hline \end{array}$
16. $\begin{array}{r} 3 \\ \times\,5 \\ \hline \end{array}$
17. $\begin{array}{r} 5 \\ \times\,1 \\ \hline \end{array}$
18. $\begin{array}{r} 6 \\ \times\,5 \\ \hline \end{array}$
19. $\begin{array}{r} 0 \\ \times\,3 \\ \hline \end{array}$
20. $\begin{array}{r} 1 \\ \times\,2 \\ \hline \end{array}$
21. $\begin{array}{r} 7 \\ \times\,0 \\ \hline \end{array}$

22. $4 \times 6 =$ _____
23. $1 \times 0 =$ _____
24. $5 \times 3 =$ _____
25. $0 \times 9 =$ _____

26. $4 \times 0 =$ _____
27. $5 \times 4 =$ _____
28. $1 \times 0 =$ _____
29. $8 \times 3 =$ _____

Mixed Review

30. $\begin{array}{r} \$6.27 \\ +\$2.66 \\ \hline \end{array}$
31. $\begin{array}{r} \$7.99 \\ -\$4.44 \\ \hline \end{array}$
32. $\begin{array}{r} \$8.31 \\ -\$5.98 \\ \hline \end{array}$
33. $\begin{array}{r} \$2.28 \\ +\$7.95 \\ \hline \end{array}$

34. $305 + 882 + 406 =$ _____
35. $761 + 75 =$ _____

36. Which shows the numbers in order from least to greatest?

    A 786   867   678

    B 867   678   786

    C 678   786   867

What is the value of the 4 in each of these numbers?

37. 9,412   38. 24   39. 46,118

_____   _____   _____

Problem Solving Strategy

Find a Pattern

Use *find a pattern* to solve.

1. Quintin's pattern is 2, 5, 8, 11, 14, and 17. What is a rule? What are the next four numbers in his pattern?

   _____

2. Vernon's pattern is 12, 15, 19, 22, and 26. What is a rule? What are the next four numbers in his pattern?

   _____

3. Laura's pattern is 14, 24, 34, 44, and 54. What is a rule? What are the next four numbers in her pattern?

   _____

4. Marianne's pattern is 31, 36, 41, 46, and 51. What is a rule? What are the next four numbers in her pattern?

   _____

5. Sharon's pattern is 54, 51, 48, 45, 42, and 39. What is a rule? What are the next four numbers in her pattern?

   _____

6. Tom's pattern is 10, 12, 13, 15, 16, and 18. What is a rule? What are the next four numbers in his pattern?

   _____

7. The first number is 4. A rule is *multiply by 2 and then subtract 3*. What are the first 6 numbers in the pattern?

   _____

8. Melinda's pattern is 9, 7, 10, 8, 11, 9, and 12. What is a rule? What are the next four numbers in her pattern?

   _____

Mixed Review

Round to the nearest thousand.

9. 7,803 _____   10. 9,975 _____   11. 9,099 _____

Write <, >, or =.

12. $5.67 _____ $5.76   13. $16.10 _____ $16.09   14. $4.89 _____ $4.90

Find 100 more than the number.

15. 2,376 _____   16. 45,903 _____   17. 119,752 _____

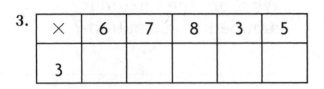

Name _____

Practice Multiplication

Complete the tables.

1.

×	3	6	7	2	5
4					

2.

×	5	4	6	7	8
5					

3.

×	6	7	8	3	5
3					

4.

×	8	2	4	3	6
2					

Find the product.

5. $1 \times 6 =$ _____

6. $2 \times 8 =$ _____

7. $2 \times 7 =$ _____

8. $4 \times 8 =$ _____

9. $3 \times 7 =$ _____

10. $4 \times 2 =$ _____

11. $8 \times 3 =$ _____

12. $4 \times 6 =$ _____

13. $2 \times 9 =$ _____

14. $4 \times 1 =$ _____

15. $5 \times 5 =$ _____

16. $1 \times 3 =$ _____

Mixed Review

17. What is the elapsed time from 11:30 P.M. to

11:45 P.M.? _____

18. $5.98
   +$2.07

19. 702
   − 67

20. $ 0.71
   +$10.49

21. 6,498
   − 3,512

22. _____ $+ 21 = 29$

23. $72 - 33 =$ _____

24. $923 + 765 =$ _____

25. $4,099 - 170 =$ _____

26. Which shows the numbers in order from greatest to least?

   A   789   897   987

   B   987   897   789

   C   897   987   789

Algebra: Missing Factors

Find the missing factor.

1. ____ $\times$ 4 = 20

2. 7 $\times$ ____ = 35

3. ____ $\times$ 6 = 18

4. 8 $\times$ ____ = 32

5. ____ $\times$ 3 = 27

6. 5 $\times$ ____ = 30

7. ____ $\times$ 5 = 15

8. ____ $\times$ 3 = 21

9. 8 $\times$ ____ = 24

10. 5 $\times$ ____ = 25

11. ____ $\times$ 4 = 24

12. ____ $\times$ 4 = 36

13. ____ $\times$ 4 = 32

14. 4 $\times$ ____ = 20

15. 2 $\times$ ____ = 12

16. 5 $\times$ ____ = 45

17. 8 $\times$ ____ = 24

18. ____ $\times$ 2 = 10

19. 3 $\times$ ____ = 27

20. ____ $\times$ 3 = 3

21. 4 $\times$ ____ = 16

22. 7 $\times$ ____ = 2 $\times$ ____

23. 5 $\times$ ____ = 45 − 5

Mixed Review

Add 8 to each.

24. 42

25. 216

26. 181

27. 437

Write the total value of each.

28. 2 dimes
3 nickels
4 pennies

29. 3 quarters
5 nickels
8 pennies

30. 3 $1-bills
4 quarters
10 dimes

31. 2 $1-bills
2 quarters
2 dimes

32. $17.25 + $6.00 = _____

33. $0.79 + $0.40 + $0.88 = _____

Complete the tables.

34.

$\times$	9	5	1	4	6
2					

35.

$\times$	4	0	3	8	7
0					

Multiply with 6

Find each product.

1. $4 \times 6 = $ 24 ~~20~~

2. $3 \times 8 = $ 24

3. $6 \times 2 = $ 12

4. $5 \times 4 = $ 20

5. $8 \times 6 = $ 48

6. $6 \times 5 = $ 30

7. $7 \times 6 = $ 30

8. $3 \times 9 = $ 27

9. $6 \times 6 = $ 36

10. $6 \times 0 = $ 0

11. $1 \times 6 = $ 6

12. $4 \times 9 = $ 36

13.
$$\begin{array}{r} 9 \\ \times 6 \\ \hline \end{array}$$

14.
$$\begin{array}{r} 7 \\ \times 4 \\ \hline \end{array}$$

15.
$$\begin{array}{r} 6 \\ \times 3 \\ \hline \end{array}$$

16.
$$\begin{array}{r} 3 \\ \times 4 \\ \hline \end{array}$$

Complete each table.

	Multiply by 2.	
17.	5	12
18.	8	16
19.	9	18

	Multiply by 6.	
20.	3	18
21.	5	30
22.	8	48

	Multiply by 4.	
23.	4	16
24.	6	12
25.	8	36

Mixed Review

Solve.

26.
$$\begin{array}{r} 4{,}009 \\ -2{,}389 \\ \hline \end{array}$$

27.
$$\begin{array}{r} 387 \\ +906 \\ \hline \end{array}$$

28.
$$\begin{array}{r} \$62.85 \\ -\$34.99 \\ \hline \end{array}$$

29.
$$\begin{array}{r} 1{,}709 \\ +\ 5{,}913 \\ \hline \end{array}$$

30.
$$\begin{array}{r} \$5.49 \\ +\$3.89 \\ \hline \end{array}$$

31.
$$\begin{array}{r} 7{,}360 \\ -2{,}507 \\ \hline \end{array}$$

32.
$$\begin{array}{r} 6{,}906 \\ -6{,}079 \\ \hline \end{array}$$

33.
$$\begin{array}{r} \$47.88 \\ +\$\ \ 6.13 \\ \hline \end{array}$$

Name _____

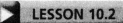

Multiply with 8

Find each product.

1. $4 \times 8 =$ _____

2. $7 \times 8 =$ _____

3. $4 \times 6 =$ _____

4. $3 \times 8 =$ _____

5. $8 \times 9 =$ _____

6. $7 \times 6 =$ _____

7. $8 \times 0 =$ _____

8. $2 \times 8 =$ _____

9. $5 \times 8 =$ _____

10. $\begin{array}{r} 7 \\ \times 2 \\ \hline \end{array}$

11. $\begin{array}{r} 1 \\ \times 8 \\ \hline \end{array}$

12. $\begin{array}{r} 8 \\ \times 6 \\ \hline \end{array}$

13. $\begin{array}{r} 8 \\ \times 8 \\ \hline \end{array}$

Complete each table.

Multiply by 5.	
14. 7	
15. 8	
16. 9	

Multiply by 6.	
17. 4	
18. 6	
19. 7	

Multiply by 8.	
20. 5	
21. 4	
22. 7	

Compare. Write $<$, $>$, or $=$ in each $\bigcirc$.

23. $8 \times 4 \bigcirc 2 \times 6$

24. $8 \times 3 \bigcirc 6 \times 8$

25. $7 \times 0 \bigcirc 8 \times 0$

26. $4 \times 5 \bigcirc 7 \times 6$

27. $8 \times 9 \bigcirc 3 \times 4$

28. $5 \times 5 \bigcirc 8 \times 8$

Mixed Review

Solve.

29. $32 + 44 + 81 =$ _____

30. $56 + 14 + 39 =$ _____

31. $82 + 8 + 18 =$ _____

32. $28 + 27 + 42 =$ _____

33. $4,290 - 3,735 =$ _____

34. $8,802 - 6,529 =$ _____

Problem Solving Skill

Use a Pictograph

For 1–3, use the pictograph.

1. Explain how to use this pictograph to find which class has the fewest students. How many students are in this class?

_____

_____

_____

_____

Third-Graders at Myra's School	
Mr. Adam's Class	♀ ♀ ♀ ♀ ♀ ♀
Miss Green's Class	♀ ♀ ♀ ♀ ♀
Mrs. Ortez's Class	♀ ♀ ♀ ♀ ♀ ♪
Mr. Hill's Class	♀ ♀ ♀ ♀ ♪
Key: Each ♀ = 4 students.	

2. How many students are in Mrs. Ortez's class?

_____

3. How many more students are in Mr. Adam's class than are in Mr. Hill's class?

_____

For 4–5, use the pictograph.

4. How many animals are in the parade?

   A 11   **C** 66
   B 64   **D** 72

5. Which numbers represented on the pictograph are multiples of 6?

   F 21 and 30   **H** 12, 21, and 30
   G 12 and 30   **J** 3, 12, and 30

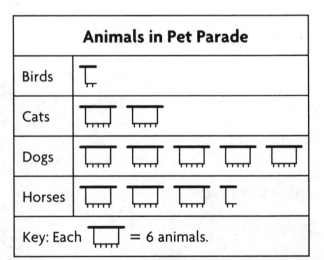

Mixed Review

Write how many there are in all.

6. 3 groups of 8   7. 7 groups of 4   8. 3 groups of 5

_____   _____   _____

© Harcourt

Name _____

Multiply with 7

Find each product.

1. $7 \times 6 =$ _____   2. $5 \times 8 =$ _____   3. $3 \times 7 =$ _____

4. $7 \times 4 =$ _____   5. $6 \times 7 =$ _____   6. $4 \times 8 =$ _____

7. $9 \times 7 =$ _____   8. $5 \times 1 =$ _____   9. $7 \times 0 =$ _____

10. $1 \times 7 =$ _____   11. $7 \times 5 =$ _____   12. $7 \times 8 =$ _____

Complete each table.

Multiply by 6.	
13.	3	
14.	7	
15.	8	

Multiply by 7.	
16.	7	
17.	9	
18.	4	

Multiply by 8.	
19.	5	
20.	9	
21.	8	

Complete.

22. $9 \times 7 =$ ____ $+ 33$ 23. $7 \times$ ____ $= 34 - 13$ 24. ____ $\times 7 = 7 + 7$

Mixed Review

Write the value of the underlined digit.

25. 53,009 _____   26. 6,842 _____   27. 92,106 _____

28. 4,222 _____   29. 11,001 _____   30. 6,681 _____

Round to the nearest hundred.

31. 5,349 _____   32. 478 _____   33. 14,780 _____

34. 26,318 _____   35. 1,159 _____   36. 879 _____

Subtract 475 from each number.

37. 690   38. 4,330   39. 2,065

_____   _____   _____

© Harcourt

Algebra: Practice the Facts

Find each product.

1. $5 \times 4 =$ _____

2. $6 \times 6 =$ _____

3. $8 \times 6 =$ _____

4. $7 \times 7 =$ _____

5. $3 \times 5 =$ _____

6. $6 \times 9 =$ _____

7. $8 \times 9 =$ _____

8. $6 \times 7 =$ _____

9. $5 \times 6 =$ _____

10. $8 \times 5 =$ _____

11. $8 \times 7 =$ _____

12. $8 \times 8 =$ _____

13. $5 \times 7 =$ _____

14. $9 \times 7 =$ _____

15. $5 \times 9 =$ _____

16.
$$\begin{array}{r} 5 \\ \times 2 \\ \hline \end{array}$$

17.
$$\begin{array}{r} 8 \\ \times 4 \\ \hline \end{array}$$

18.
$$\begin{array}{r} 7 \\ \times 8 \\ \hline \end{array}$$

19.
$$\begin{array}{r} 7 \\ \times 6 \\ \hline \end{array}$$

20.
$$\begin{array}{r} 9 \\ \times 8 \\ \hline \end{array}$$

21.
$$\begin{array}{r} 4 \\ \times 4 \\ \hline \end{array}$$

22.
$$\begin{array}{r} 9 \\ \times 3 \\ \hline \end{array}$$

23.
$$\begin{array}{r} 4 \\ \times 7 \\ \hline \end{array}$$

Find each missing factor.

24. $5 \times$ _____ $= 45$

25. $9 \times$ _____ $= 36$

26. $8 \times$ _____ $= 16$

27. $3 \times$ _____ $= 27$

28. $7 \times$ _____ $= 63$

29. _____ $\times 8 = 24$

30. _____ $\times 6 = 54$

31. _____ $\times 4 = 28$

32. $6 \times$ _____ $= 24$

Mixed Review

Add.

33.
$$\begin{array}{r} 45 \\ 16 \\ +27 \\ \hline \end{array}$$

34.
$$\begin{array}{r} 43 \\ 57 \\ +87 \\ \hline \end{array}$$

35.
$$\begin{array}{r} 44 \\ 55 \\ +66 \\ \hline \end{array}$$

36.
$$\begin{array}{r} 73 \\ 64 \\ 46 \\ +11 \\ \hline \end{array}$$

Multiply with 9 and 10

Find the product.

1. $\begin{array}{r} 9 \\ \times\,5 \\ \hline \end{array}$
2. $\begin{array}{r} 10 \\ \times\,9 \\ \hline \end{array}$
3. $\begin{array}{r} 10 \\ \times\,6 \\ \hline \end{array}$
4. $\begin{array}{r} 10 \\ \times\,8 \\ \hline \end{array}$
5. $\begin{array}{r} 9 \\ \times\,4 \\ \hline \end{array}$

6. $\begin{array}{r} 9 \\ \times\,6 \\ \hline \end{array}$
7. $\begin{array}{r} 10 \\ \times\,5 \\ \hline \end{array}$
8. $\begin{array}{r} 10 \\ \times\,3 \\ \hline \end{array}$
9. $\begin{array}{r} 7 \\ \times\,9 \\ \hline \end{array}$
10. $\begin{array}{r} 10 \\ \times\,2 \\ \hline \end{array}$

11. $\begin{array}{r} 9 \\ \times\,3 \\ \hline \end{array}$
12. $\begin{array}{r} 10 \\ \times\,4 \\ \hline \end{array}$
13. $\begin{array}{r} 9 \\ \times\,9 \\ \hline \end{array}$
14. $\begin{array}{r} 10 \\ \times\,7 \\ \hline \end{array}$
15. $\begin{array}{r} 8 \\ \times\,9 \\ \hline \end{array}$

16. $8 \times 10 =$ _____
17. $9 \times 2 =$ _____
18. $1 \times 10 =$ _____

19. $1 \times 9 =$ _____
20. $9 \times 10 =$ _____
21. $9 \times 5 =$ _____

22. $10 \times 2 =$ _____
23. $10 \times 8 =$ _____
24. $9 \times 7 =$ _____

Find the missing factor.

25. _____ $\times 8 = 0$
26. _____ $\times 2 = 20$
27. $7 \times$ _____ $= 7$

28. $9 \times$ _____ $= 6 \times 3$
29. $5 \times 8 =$ _____ $\times 10$
30. _____ $\times 9 = 6 \times 6$

Complete each table.

Multiply by 9	
31.	9	
32.	8	

Multiply by 7	
33.	6	
34.	8	

Multiply by 10	
35.	7	
36.	9	

Mixed Review

Add or subtract.

37. $\begin{array}{r} \$8.09 \\ -\$3.55 \\ \hline \end{array}$
38. $\begin{array}{r} \$7.00 \\ -\$6.99 \\ \hline \end{array}$
39. $\begin{array}{r} \$5.55 \\ \$4.44 \\ +\$3.33 \\ \hline \end{array}$
40. $\begin{array}{r} \$1.29 \\ \$1.39 \\ +\$1.49 \\ \hline \end{array}$

Algebra: Find a Rule

Write a rule for each table. Then complete the table.

1.

Flutes	2	3	4	5	6
Trumpets	6	9	12		

Rule: _____

2.

Cups	1	2	3	4	5	6
Ounces	8	16	24			

Rule: _____

3.

Plates	5	6	7	8	9	10
Bowls	10	12	14	16		

Rule: _____

4.

Plants	4	5	6	7	8	9
Flowers	24	30	36			

Rule: _____

5. Each box holds 4 toys. How many toys do 5 boxes hold?

Boxes	1	2			
Toys	4	8			

Rule: _____

6. Four shelves hold 36 toys. How many toys do 9 shelves hold?

Shelves	4	5	6			
Toys	36	45				

Rule: _____

Mixed Review

Find the elapsed time.

7. 7:00 P.M. to 8:30 P.M.

_____

8. 4:00 A.M. to noon

_____

9. 9:00 A.M. to 1:00 P.M.

_____

10. 6:30 P.M. to 10:15 P.M.

_____

Use mental math to find the sum.

11.
$$\begin{array}{r} 52 \\ 48 \\ 24 \\ + 26 \\ \hline \end{array}$$

12.
$$\begin{array}{r} 17 \\ 13 \\ 16 \\ + 14 \\ \hline \end{array}$$

13.
$$\begin{array}{r} 51 \\ 49 \\ 47 \\ + 53 \\ \hline \end{array}$$

14.
$$\begin{array}{r} 19 \\ 21 \\ 15 \\ + 15 \\ \hline \end{array}$$

Name _____

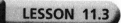

Algebra: Multiply with 3 Factors

Find each product.

1. $(3 \times 2) \times 3 =$ _____   **2.** $6 \times (4 \times 2) =$ _____   **3.** $(3 \times 3) \times 5 =$ _____

4. $(2 \times 2) \times 8 =$ _____   **5.** $(1 \times 4) \times 7 =$ _____   **6.** $4 \times (7 \times 1) =$ _____

7. $6 \times (0 \times 7) =$ _____   **8.** $(3 \times 3) \times 10 =$ _____   **9.** $(7 \times 1) \times 8 =$ _____

Use parentheses. Find the product.

10. $3 \times 3 \times 6 =$ _____   **11.** $4 \times 4 \times 2 =$ _____   **12.** $9 \times 3 \times 2 =$ _____

13. $7 \times 2 \times 2 =$ _____   **14.** $2 \times 4 \times 7 =$ _____   **15.** $4 \times 9 \times 1 =$ _____

16. $4 \times 2 \times 5 =$ _____   **17.** $3 \times 2 \times 10 =$ _____   **18.** $4 \times 2 \times 7 =$ _____

Find the missing factor.

19. $(8 \times$ _____$) \times 8 = 0$ **20.** _____ $\times (3 \times 2) = 36$ **21.** $($_____ $\times 4) \times 3 = 12$

22. $6 \times (3 \times$ _____$) = 54$ **23.** $(3 \times 3) \times$ _____ $= 90$ **24.** _____ $\times (5 \times 2) = 80$

25. $($_____ $\times 1) \times 1 = 6$ **26.** $4 \times ($_____ $\times 4) = 32$ **27.** $(2 \times 4) \times$ _____ $= 64$

Mixed Review

Write the missing number that makes each sentence true.

28. $9 +$ _____ $= 20$ **29.** $8 =$ _____ $- 3$

30. _____ $+ 13 = 44$ **31.** $560 = 200 +$ _____

Write $<$, $>$, or $=$ for each $\bigcirc$.

32. $544 \bigcirc 544$ **33.** $5,106 \bigcirc 5,099$ **34.** $467 + 3 \bigcirc 471$

Complete the pattern.

35. 6, 12, 18, 24, _____, _____, _____, _____

36. 39, 49, _____, 69, _____, _____, _____

37. 75, 70, 65, 60, 55, _____, _____, _____

Multiplication Properties

Find the product. Tell which property you used to help you.

1. $8 \times 7 =$ _____

2. $1 \times 6 =$ _____

3. $(2 \times 3) \times 4 =$ _____

_____

_____

_____

4. $7 \times 0 =$ _____

5. $5 \times (2 \times 4) =$ _____

6. $9 \times 1 =$ _____

_____

_____

_____

7. $9 \times 8 =$ _____

8. $(2 \times 6) \times 3 =$ _____

9. $0 \times 4 =$ _____

_____

_____

_____

10. $1 \times 5 =$ _____

11. $8 \times 0 =$ _____

12. $7 \times 6 =$ _____

_____

_____

_____

Write the missing number.

13. $4 \times 3 =$ _____ $\times 4$

14. $5 \times 9 = (5 \times 3) + (5 \times$ _____ $)$

15. $3 \times (2 \times 6) = (3 \times$ _____ $) \times 6$

16. $(8 \times 2) \times 4 =$ _____ $\times (2 \times 4)$

17. _____ $\times 9 = 9 \times 6$

18. $4 \times 7 = ($ _____ $\times 5) + ($ _____ $\times 2)$

Mixed Review

Solve.

19. $\$4.57$ $+ \$7.39$	**20.** $\$9.03$ $- \$2.54$	**21.** $\$26.88$ $+ \$75.42$	**22.** $\$50.00$ $- \$24.99$

Round each number to the nearest thousand.

23. $2,463$ _____

24. $8,711$ _____

25. 932 _____

26. $4,300$ _____

27. $6,514$ _____

28. $7,820$ _____

Problem Solving Skill
Multistep Problems

Solve.

1. Taylor bought 6 used books that cost $2 each. He also bought 3 used books that cost $4 each. How much did Taylor spend on used books?

_____

2. Tina has 3 rows of 8 rocks in her rock collection. She wants to double her collection. How many rocks will Tina have when she doubles her collection?

_____

3. Howard has $138 and Tess has $149. They need a total of $250 to buy a recliner chair for their father. How much more money do they have than they need?

_____

4. To raise money for school, Megan sold 8 magazine subscriptions. Parker sold 7 subscriptions. Each subscription raises $5 for the school. How much money did they raise in all?

_____

5. The Romers drove 613 miles in 3 days. They drove 251 miles the first day and 168 miles the second day. How far did they drive on the third day?

_____

6. Two friends are comparing money. Bert has 8 quarters and 7 dimes. Ernie has 10 quarters and 7 nickels. Who has the most money? How much more money than his friend does he have?

_____

Mixed Review

Continue the pattern.

7. 20, 40, 60, 80, _?_, _?_, _?_

_____

8. 12, 14, 15, 17, 18, 20, _?_, _?_

_____

Find the product.

9. $(2 \times 3) \times 9 =$ _____

10. $6 \times (3 \times 3) =$ _____

Meaning of Division

Complete the table. Use counters to help.

	Counters	Number of equal groups	Number in each group
1.	10	2	
2.	12		6
3.	16	4	
4.	18		6
5.	21	3	

For 6–9, use counters.

6. Four family members want to share a bag of 20 pretzels equally. How many pretzels will each person get?

_____

7. Carrie and two friends are sharing a pizza cut into 12 slices. If each person eats the same number of slices, how many slices will each person get?

_____

8. Six students are sharing the job of watering the classroom plants. Each student waters 3 plants. How many plants are in the classroom altogether?

_____

9. Emma's friends are helping her write a total of 16 invitations. Each person has 4 invitations to write. How many people are working together?

_____

Mixed Review

Solve.

10. $77.42
    −$24.59

11. 3,071
    + 809

12. 468
    −312

13. 818
    −607

14. 6
    ×5

15. 8
    ×9

16. 7
    ×4

17. 3
    ×2

Subtraction and Division

Write a division sentence for each.

1.

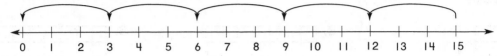

_____

2.

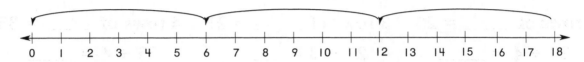

_____

3.
$$\frac{10}{-2} \Big/ \frac{8}{-2} \Big/ \frac{6}{-2} \Big/ \frac{4}{-2} \Big/ \frac{2}{-2}$$
$$\;\;\;8\quad\;\;6\quad\;\;4\quad\;\;2\quad\;\;0$$

_____

4.
$$\frac{16}{-4} \Big/ \frac{12}{-4} \Big/ \frac{8}{-4} \Big/ \frac{4}{-4}$$
$$\;\;12\quad\;\;8\quad\;\;4\quad\;\;0$$

_____

Use a number line or subtraction to solve.

5. $12 \div 3 =$ _____

6. $20 \div 4 =$ _____

7. $30 \div 5 =$ _____

8. $6 \div 2 =$ _____

_____   _____

Mixed Review

9.
$$\begin{array}{r} 271 \\ +409 \\ \hline \end{array}$$

10.
$$\begin{array}{r} 9{,}006 \\ -7{,}847 \\ \hline \end{array}$$

11.
$$\begin{array}{r} 7 \\ \times 6 \\ \hline \end{array}$$

12.
$$\begin{array}{r} 4 \\ \times 9 \\ \hline \end{array}$$

13. $7 \times 7 =$ _____

14. $8 \times 3 =$ _____

15. $8 \times 6 =$ _____

Algebra: Multiplication and Division

Complete.

1.
4 rows of _____ = 20

$20 \div 4 =$ _____

2.
3 rows of _____ = 21

$21 \div 3 =$ _____

3.
4 rows of _____ = 36

$36 \div 4 =$ _____

Complete each number sentence. Draw an array to help.

4. $6 \times$ _____ $= 18$

5. $32 \div 8 =$ _____

6. $4 \times 5 =$ _____

Find the number that the variable stands for.

7. $2 \times p = 10$

$p =$ _____

8. $q \times 7 = 21$

$q =$ _____

9. $8 \times r = 16$

$r =$ _____

Mixed Review

10.
$$\begin{array}{r} 760 \\ -152 \\ \hline \end{array}$$

11.
$$\begin{array}{r} 3{,}789 \\ +\ 534 \\ \hline \end{array}$$

12.
$$\begin{array}{r} 8{,}117 \\ -5{,}833 \\ \hline \end{array}$$

13.
$$\begin{array}{r} 6{,}211 \\ -5{,}819 \\ \hline \end{array}$$

14.
$$\begin{array}{r} 380 \\ +8{,}495 \\ \hline \end{array}$$

15.
$$\begin{array}{r} 7{,}117 \\ +2{,}981 \\ \hline \end{array}$$

© Harcourt

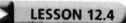

Algebra: Fact Families

Write the fact family.

1. 4, 9, 36

2. 8, 3, 24

3. 6, 4, 24

_____

_____

_____

_____

_____

_____

_____

_____

_____

_____

_____

_____

4. 6, 6, 36

5. 7, 7, 49

6. 5, 5, 25

_____

_____

_____

_____

_____

_____

Find the quotient or product.

7. $5 \times 7 =$ _____ **8.** $7 \times 5 =$ _____ **9.** $35 \div 7 =$ _____ **10.** $35 \div 5 =$ _____

Write the other three sentences in the fact family.

11. $6 \times 3 = 18$

12. $4 \times 5 = 20$

13. $2 \times 7 = 14$

_____

_____

_____

_____

_____

_____

_____

_____

_____

Mixed Review

Write $+$, $-$, $\times$, or $\div$ in each $\bigcirc$.

14. $36 \bigcirc 4 = 9$

15. $18 \bigcirc 12 = 6$

16. $2 \bigcirc 8 = 16$

17. $72 \bigcirc 9 = 8$

18. $14 \bigcirc 4 = 10$

19. $9 \bigcirc 6 = 54$

Problem Solving Strategy

Write a Number Sentence

Write a number sentence to solve.

1. Mrs. Scott bought 3 packages of hot dogs. Each package has 8 hot dogs. How many hot dogs did she buy in all?

_____

2. A class of 27 students is working in groups of 3 on an art project. How many groups are there?

_____

3. Melissa took 24 photographs. She put 4 photographs on each page of her album. How many pages did she use?

_____

4. Tim planted 5 rows of corn. There are 6 corn plants in each row. How many corn plants are there in all?

_____

Mixed Review

5. $2.42
   +$5.65

6. $4.91
   −$0.76

7. $8.56
   −$3.28

8. $7.99
   +$1.99

9.   8
   × 5

10. 5
   × 8

11. 9
   × 9

12. 6
   × 8

13. $3 \times 7 =$ _____   14. $6 \times 9 =$ _____   15. $10 \times 4 =$ _____   16. $4 \times 7 =$ _____

Write $+, -, \times,$ or $\div$ in each $\bigcirc$.

17. $84 \bigcirc 25 = 59$

18. $6 \bigcirc 8 = 48$

19. $32 \bigcirc 73 = 105$

20. $54 \bigcirc 9 = 63$

21. $7 \bigcirc 6 = 42$

22. $9 \bigcirc 5 = 45$

Divide by 2 and 5

Find each missing factor or quotient.

1. $2 \times$ _____ $= 8$
2. $30 \div 5 =$ _____
3. $16 \div 2 =$ _____

4. $45 \div 5 =$ _____
5. $5 \times$ _____ $= 25$
6. $8 \div 2 =$ _____

7. $5 \times$ _____ $= 15$
8. $2 \times$ _____ $= 20$
9. $2 \times$ _____ $= 12$

Find each quotient.

10. $18 \div 2 =$ _____
11. $35 \div 5 =$ _____
12. $40 \div 5 =$ _____

13. $4 \div 2 =$ _____
14. $10 \div 2 =$ _____
15. $5 \div 5 =$ _____

16. $5\overline{)30}$
17. $2\overline{)14}$
18. $5\overline{)20}$
19. $5\overline{)5}$

20. $2\overline{)12}$
21. $2\overline{)8}$
22. $5\overline{)15}$
23. $5\overline{)40}$

Complete.

24. $20 \div 2 =$ _____ $+ 6$
25. $15 \div 5 =$ _____ $\times 1$
26. $40 \div 5 =$ _____ $\times 2$

Mixed Review

Solve.

27. $9 \times 3 \times$ _____ $= 81$
28. _____ $\times 6 \times 2 = 12$
29. $9 \times$ _____ $= 63$

Add 1,000 to each.

30. 32,605
31. 20,001
32. 518
33. 6

_____ _____ _____ _____

Write A.M. or P.M.

34. ten minutes after midnight
35. time to go to bed
36. ten minutes before noon
37. ten minutes before midnight

_____ _____ _____ _____

Name _____

Divide by 3 and 4

Write the multiplication fact you can use to find the
quotient. Then write the quotient.

1. $36 \div 4$ **2.** $21 \div 3$ **3.** $28 \div 4$

_____ _____ _____

_____ _____ _____

Find each quotient.

4. $18 \div 3 =$ _____ **5.** $32 \div 4 =$ _____ **6.** $30 \div 3 =$ _____

7. $8 \div 2 =$ _____ **8.** $12 \div 3 =$ _____ **9.** $12 \div 4 =$ _____

10. $3\overline{)15}$ **11.** $4\overline{)28}$ **12.** $3\overline{)27}$ **13.** $4\overline{)16}$

14. $4\overline{)32}$ **15.** $3\overline{)9}$ **16.** $4\overline{)8}$ **17.** $3\overline{)30}$

Complete.

18. $12 \div 4 =$ _____ $\times 3$ **19.** $24 \div 4 =$ _____ $\times 3$ **20.** $27 \div 3 =$ _____ $\times 3$

Mixed Review

Solve.

21. $\begin{array}{r} 8 \\ \times 9 \\ \hline \end{array}$ **22.** $\begin{array}{r} 7 \\ \times 8 \\ \hline \end{array}$ **23.** $\begin{array}{r} 6 \\ \times 7 \\ \hline \end{array}$ **24.** $\begin{array}{r} 5 \\ \times 6 \\ \hline \end{array}$ **25.** $\begin{array}{r} 4 \\ \times 5 \\ \hline \end{array}$

26. $\begin{array}{r} 9 \\ \times 9 \\ \hline \end{array}$ **27.** $\begin{array}{r} 8 \\ \times 8 \\ \hline \end{array}$ **28.** $\begin{array}{r} 7 \\ \times 7 \\ \hline \end{array}$ **29.** $\begin{array}{r} 6 \\ \times 6 \\ \hline \end{array}$ **30.** $\begin{array}{r} 5 \\ \times 5 \\ \hline \end{array}$

31. $\begin{array}{r} \$13.87 \\ + \$25.62 \\ \hline \end{array}$ **32.** $\begin{array}{r} \$45.16 \\ + \$82.37 \\ \hline \end{array}$ **33.** $\begin{array}{r} \$63.27 \\ + \$37.92 \\ \hline \end{array}$ **34.** $\begin{array}{r} \$49.95 \\ + \$77.85 \\ \hline \end{array}$

Divide with 1 and 0

Find each quotient.

1. $7 \div 7 =$ _____
2. $0 \div 5 =$ _____
3. $4 \div 1 =$ _____

4. $8 \div 1 =$ _____
5. $6 \div 6 =$ _____
6. $0 \div 3 =$ _____

7. $2 \div 2 =$ _____
8. $0 \div 8 =$ _____
9. $2 \div 1 =$ _____

10. $0 \div 4 =$ _____
11. $3 \div 1 =$ _____
12. $5 \div 5 =$ _____

13. $4 \div 4 =$ _____
14. $9 \div 1 =$ _____
15. $0 \div 2 =$ _____

16. $7 \div 1 =$ _____
17. $9 \div 9 =$ _____
18. $6 \div 1 =$ _____

19. $0 \div 1 =$ _____
20. $0 \div 9 =$ _____
21. $3 \div 3 =$ _____

Compare. Write $<$, $>$, or $=$ for each $\bigcirc$.

22. $7 \div 7 \bigcirc 7 \div 1$
23. $9 \div 9 \bigcirc 10 - 9$
24. $5 \div 1 \bigcirc 5 + 1$

25. $0 \div 6 \bigcirc 6 + 0$
26. $2 + 4 \bigcirc 0 \div 6$
27. $3 \div 1 \bigcirc 3 \times 1$

Mixed Review

Solve.

28.
$$\begin{array}{r} 475 \\ - 352 \\ \hline \end{array}$$

29.
$$\begin{array}{r} 450 \\ + 640 \\ \hline \end{array}$$

30.
$$\begin{array}{r} 7,991 \\ - 4,328 \\ \hline \end{array}$$

31.
$$\begin{array}{r} 665 \\ + 392 \\ \hline \end{array}$$

32.
$$\begin{array}{r} \$3.67 \\ + \$2.33 \\ \hline \end{array}$$

33.
$$\begin{array}{r} \$4.27 \\ + \$3.59 \\ \hline \end{array}$$

34.
$$\begin{array}{r} \$28.95 \\ - \$17.60 \\ \hline \end{array}$$

35.
$$\begin{array}{r} \$13.40 \\ - \$11.72 \\ \hline \end{array}$$

Find each missing number.

36. $6 \div$ _____ $= 2$
37. $8 \div$ _____ $= 4$

38. _____ $\div 4 = 1$
39. _____ $\div 7 = 3$

Algebra: Expressions and Equations

Write an expression to describe each problem.

1. Kim has 18 craft sticks. His mother gives him 3 more. How many craft sticks does he have now?

2. Four students share 36 tacks. How many tacks does each student get?

_____

3. Beth has an album with 9 pages. She can fit 8 photos on each page. How many photos can be in the album?

4. Tim stacked 20 blocks. He then took away 8 of them. How many blocks remained in the stack?

_____

Write an equation to solve.

5. Vinnie is 5 years younger than Carly. Vinnie is 15 years old. How old is Carly?

6. Mindy has $1.00. She spends $0.85 on lunch. How much money does she have left?

_____

7. Pauline has 35 baseball cards. She buys 5 more cards. How many cards does she have altogether?

8. Matthew is 2 times as old as Greg. Greg is 6 years old. How old is Matthew?

_____

Mixed Review

Add, subtract, multiply, or divide.

9. $\begin{array}{r} 6 \\ \times 3 \\ \hline \end{array}$

10. $\begin{array}{r} 45 \\ +68 \\ \hline \end{array}$

11. $\begin{array}{r} 101 \\ -73 \\ \hline \end{array}$

12. $5\overline{)45}$

Write the missing number in each problem.

13. $\begin{array}{r} 3,672 \\ + \\ \hline 4,020 \end{array}$

14. $\begin{array}{r} 888 \\ - \\ \hline 323 \end{array}$

15. $\begin{array}{r} 4 \\ \times \\ \hline 36 \end{array}$

16. $9\overline{)}^{4}$

Problem Solving Skill

Choose the Operation

Choose the operation. Write an equation.
Then solve.

1. There are 9 mice in each cage. There are 3 cages. How many mice are there in all?

   _____

   _____

2. Izzy and Tom are cats. Izzy weighs 9 pounds and Tom weighs 12 pounds. How much more does Tom weigh than Izzy?

   _____

   _____

3. Mrs. Ellis buys 9 cans of cat food. She already has 8 cans of cat food at home. How many cans does she have now?

   _____

   _____

4. Mr. Davis has 24 goldfish. He puts 8 fish in each fish bowl. How many fish bowls does he use?

   _____

   _____

Mixed Review

5. $0 \div 3 =$ _____

6. $18 \div 2 =$ _____

7. $42 + 39 + 72 =$ _____

8. $742 - 329 =$ _____

9. Divide 30 by 3. _____

10. Divide 36 by 4. _____

11. $\begin{array}{r} 4,422 \\ -\ 3,795 \\ \hline \end{array}$

12. $\begin{array}{r} 6,219 \\ -\ 1,706 \\ \hline \end{array}$

13. $\begin{array}{r} 3,290 \\ +2,416 \\ \hline \end{array}$

14. $\begin{array}{r} 5,554 \\ -\ 4,787 \\ \hline \end{array}$

Find each missing factor, divisor, or quotient.

15. _____ $\times\ 4 = 24$

16. $49 \div$ _____ $= 7$

17. $35 \div 5 =$ _____

18. $8 \times$ _____ $= 64$

Divide by 6, 7, and 8

Find the missing factor and quotient.

1. $6 \times$ _____ $= 30$ $30 \div 6 =$ _____

2. $8 \times$ _____ $= 56$ $56 \div 8 =$ _____

3. $7 \times$ _____ $= 63$ $63 \div 7 =$ _____

Find the quotient.

4. $18 \div 6 =$ _____   5. $32 \div 8 =$ _____   6. $40 \div 8 =$ _____

7. $49 \div 7 =$ _____   8. $12 \div 6 =$ _____   9. $35 \div 7 =$ _____

10. $7\overline{)14}$ 11. $7\overline{)28}$ 12. $6\overline{)24}$ 13. $7\overline{)56}$

14. $7\overline{)63}$ 15. $6\overline{)30}$ 16. $6\overline{)54}$ 17. $8\overline{)24}$

Complete.

18. $36 \div 6 =$ _____ $\times 3$ 19. $56 \div 7 =$ _____ $+ 3$ 20. $8 \div 8 =$ _____ $- 3$

Mixed Review

Write the numbers in order from greatest to least.

21.	22.	23.	24.
19	2,013	315	30,500
43	2,130	272	30,099
38	3,120	156	30,122

_____   _____   _____   _____

_____   _____   _____   _____

Add.

25.	26.	27.	28.	29.
14	74	411	7,000	6,100
22	28	260	3,000	5,100
+ 69	+ 32	+ 591	+ 1,000	+ 3,000

Divide by 9 and 10

Complete each table.

1.
÷	18	27	36	45
9				

2.
÷	30	40	50	60
10				

Find the quotient.

3. $72 \div 9 =$ _____   **4.** $63 \div 9 =$ _____   **5.** $40 \div 8 =$ _____

6. $60 \div 10 =$ _____   **7.** $9 \div 1 =$ _____   **8.** $81 \div 9 =$ _____

9. $10 \overline{)10}$ **10.** $9 \overline{)27}$ **11.** $9 \overline{)54}$ **12.** $10 \overline{)70}$

13. $9 \overline{)63}$ **14.** $9 \overline{)90}$ **15.** $10 \overline{)90}$ **16.** $10 \overline{)100}$

Complete.

17. $54 \div 9 =$ _____ $\times 3$ **18.** $80 \div 10 =$ _____ $- 7$ **19.** $36 \div 9 =$ _____ $+ 3$

Write $+$, $-$, $\times$, or $\div$ for each $\bigcirc$.

20. $36 \bigcirc 4 = 9$ **21.** $18 \bigcirc 6 = 12$

22. $9 \bigcirc 3 = 27$ **23.** $16 \bigcirc 8 = 24$

Mixed Review

Solve.

24. Divide 45 by 5.   **25.** Divide 24 by 6.   **26.** Divide 48 by 8.

_____   _____   _____

Write the time.

27. 18 minutes after noon   **28.** 18 minutes before noon   **29.** 20 minutes before 1:15 P.M.

_____   _____   _____

Practice Division Facts

Write a division sentence for each.

1. @ @   @ @
   @ @   @ @

2. @ @ @ @ @ @ @
   @ @ @ @ @ @ @
   @ @ @ @ @ @ @

3. $\begin{array}{r} 20 \\ -10 \\ \hline 10 \end{array}$ ↗ $\begin{array}{r} 10 \\ -10 \\ \hline 0 \end{array}$

_____   _____   _____

Find the missing factor and quotient.

4. $7 \times$ _____ $= 49$ $49 \div 7 =$ _____

5. $6 \times$ _____ $= 54$ $54 \div 6 =$ _____

Find the quotient.

6. $36 \div 6 =$ _____   7. $24 \div 8 =$ _____   8. $42 \div 7 =$ _____

9. $56 \div 8 =$ _____   10. $63 \div 7 =$ _____   11. $14 \div 2 =$ _____

12. $8\overline{)64}$ 13. $10\overline{)10}$ 14. $5\overline{)35}$ 15. $9\overline{)27}$

16. $7\overline{)70}$ 17. $5\overline{)30}$ 18. $4\overline{)36}$ 19. $7\overline{)49}$

Compare. Write $<$, $>$, or $=$ for each $\bigcirc$.

20. $36 - 6 \bigcirc 8 \times 3$ 21. $18 \div 9 \bigcirc 0 + 3$ 22. $64 \div 8 \bigcirc 2 \times 4$

Mixed Review

Write a multiplication sentence for each.

23. @ @ @
    @ @ @
    @ @ @

24. @ @
    @ @
    @ @
    @ @

25. @ @ @ @
    @ @ @ @
    @ @ @ @
    @ @ @ @
    @ @ @ @

26. @ @ @
    @ @ @
    @ @ @
    @ @ @
    @ @ @
    @ @ @

_____   _____   _____   _____

Algebra: Find the Cost

Complete the table. Use the price list at the right.

1.

Hot dogs	2	4	6	8	10
Cost					

Lunch To Go	
Tuna salad	$5
Soft drink	$1
Hot dog	$2
Hamburger	$4

For 2–10, use the price list at the right to find the cost of each number of items.

2. 5 soft drinks

_____

3. 8 hamburgers

_____

4. 9 tuna salads

_____

5. 7 tuna salads

_____

6. 5 hot dogs

_____

7. 6 hamburgers

_____

8. 9 hot dogs

_____

9. 3 soft drinks

_____

10. 5 tuna salads

_____

Find the cost of one of each item.

11. 6 pens cost $18.

_____

12. 4 CDs cost $36.

_____

13. 9 salads cost $36.

_____

14. 8 mice cost $40.

_____

15. 7 gerbils cost $56.

_____

16. 9 hamsters cost $45.

_____

17. 3 cages cost $30.

_____

18. 8 balls cost $48.

_____

19. 5 games cost $35.

_____

Mixed Review

Continue each pattern.

20. 3, 10, 13, 20, 23, 30, _____, _____

21. 9, 7, 10, 8, 11, 9, _____, _____

Add.

22.
$$\begin{array}{r} 1,382 \\ 7,344 \\ +\ 2,196 \\ \hline \end{array}$$

23.
$$\begin{array}{r} 1,152 \\ 634 \\ +\ 776 \\ \hline \end{array}$$

24.
$$\begin{array}{r} 4,848 \\ 7,474 \\ +\ 4,994 \\ \hline \end{array}$$

25.
$$\begin{array}{r} 618 \\ 554 \\ +\ 920 \\ \hline \end{array}$$

Problem Solving Strategy

Work Backward

Work backward to solve.

1. Mr. Ruiz sells mailboxes. He sold 5 mailboxes and then made 12 more. Now he has 15 mailboxes. How many did he begin with?

_____

2. Paul has 23 outfielders and 19 pitchers in his baseball card collection. If he has a total of 95 cards, how many are not outfielders or pitchers?

_____

3. Josh has 17 quarters and 28 dimes in his bank. There are 102 coins in the bank. How many are not quarters or dimes?

_____

4. Tim sells picture frames. He sold 14 and then made 8 more. Now he has 23 frames. How many did he begin with?

_____

Mixed Review

Solve.

5.	6.	7.	8.
274	$1.92	$2.52	381
36	$3.34	$1.12	77
+183	+$0.57	+$0.67	+342

Continue each pattern.

9. 2, 9, 16, 23, _____, _____

10. 36, 31, 26, 21, _____, _____

11. 11, 14, 17, 20, _____, _____

12. 64, 58, 52, 46, _____, _____

Multiply.

13. $9 \times 10 =$ _____

14. $7 \times 4 =$ _____

15. $8 \times 8 =$ _____

16. $4 \times 3 =$ _____

17. $5 \times 9 =$ _____

18. $7 \times 5 =$ _____

19. $6 \times 7 =$ _____

20. $9 \times 7 =$ _____

Collect Data

1. Make a tally table of four kinds of pets. Ask some of your classmates which pet they like best. Make a tally mark beside the name of the pet each one chooses.

2. Use the data from your tally table to make a frequency table.

3. Which type of pet did the most classmates choose? the fewest?

_____

4. Compare your tables with those of your classmates. Did everyone get the same results?

_____

FAVORITE PETS	
Name	**Tally**

FAVORITE PETS	
Name	**Number**

Mixed Review

Write $<$, $>$, or $=$ for each $\bigcirc$.

5. $6 \div 1 \bigcirc 6 \div 6$

6. $10 \times 4 \bigcirc 5 \times 9$

7. $12 + 12 \bigcirc 10 + 13$

8. $354 \bigcirc 370 - 30$

9. $236 + 3 \bigcirc 239$

10. $54 \div 9 \bigcirc 70 \div 10$

11. $3 \times 3 \bigcirc 10 \times 1$

12. $0 \div 6 \bigcirc 0 \div 7$

Solve.

13. $\begin{array}{r} 500 \\ -\ 238 \\ \hline \end{array}$

14. $\begin{array}{r} 104 \\ -\ 57 \\ \hline \end{array}$

15. $\begin{array}{r} 78 \\ +\ 46 \\ \hline \end{array}$

16. $\begin{array}{r} 518 \\ +\ 203 \\ \hline \end{array}$

17. $\begin{array}{r} 729 \\ +\ 819 \\ \hline \end{array}$

Use Data from a Survey

For 1–4, use the tally table.

1. List the games in order from the most to the least chosen.

   _____

   _____

   _____

OUR FAVORITE GAMES	
Game	Tally
Follow-the-Leader	$\cancel{||||}\ ||$
Jump Rope	$\cancel{||||}\ \cancel{||||}\ \cancel{||||}\ |$
Tether Ball	$\cancel{||||}\ \cancel{||||}\ |$
Four-Square	$||||$

2. How many people answered the survey?

   _____

3. How many more people like jump rope than four-square?

   _____

4. How many fewer people like follow-the-leader than jump rope?

   _____

Mixed Review

5. $\begin{array}{r} 106 \\ +\ 894 \\ \hline \end{array}$

6. $\begin{array}{r} 1,219 \\ +\ 6,537 \\ \hline \end{array}$

7. $\begin{array}{r} 9,213 \\ -\ 3,219 \\ \hline \end{array}$

8. $\begin{array}{r} 4,266 \\ -\ 875 \\ \hline \end{array}$

9. $\begin{array}{r} 8 \\ \times\ 4 \\ \hline \end{array}$

10. $\begin{array}{r} 1 \\ \times\ 9 \\ \hline \end{array}$

11. $\begin{array}{r} 12 \\ \times\ 0 \\ \hline \end{array}$

12. $\begin{array}{r} 4 \\ \times\ 6 \\ \hline \end{array}$

13. $\begin{array}{r} 7 \\ \times\ 7 \\ \hline \end{array}$

14. Find the sum of 804 and 159. _____

15. Which number is greater: 6,232 or 6,323? _____

16. Round 2,975 to the nearest thousand. _____

Classify Data

For 1–5, use the table.

1. How many dogs have short, brown hair?

   _____

2. How many dogs have medium hair?

   _____

3. How many dogs have white hair?

   _____

DOGS OWNED BY STUDENTS				
	Black Hair	White Hair	Brown Hair	Golden Hair
Short Hair	3	4	1	3
Medium Hair	2	2	0	1
Long Hair	1	3	3	2

4. What color hair do only 4 dogs have?

   _____

5. How many dogs are owned by the class?

   _____

6. Look at the marbles at the right. Make a table to classify, or group, the marbles.

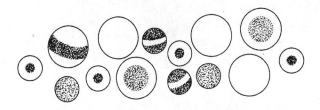

Mixed Review

Solve.

7. $\begin{array}{r} 7{,}004 \\ +1{,}664 \\ \hline \end{array}$
8. $\begin{array}{r} 1{,}241 \\ -1{,}123 \\ \hline \end{array}$
9. $\begin{array}{r} 3{,}536 \\ +5{,}544 \\ \hline \end{array}$
10. $\begin{array}{r} 9{,}432 \\ -6{,}780 \\ \hline \end{array}$

Problem Solving Strategy

Make a Table

Solve.

1. Karen and José are using the spinner and coin shown below. They spin the spinner and flip the coin. Then they record the results. They repeat this 15 times. Show how they could organize the data in a table.

2. Phillip is using the two coins shown below. He will toss both coins 25 times and record the results after each pair of tosses. Show how he could organize the data in a table.

Mixed Review

Round to the nearest 100 and 1,000.

3. 1,355 _____

4. 5,667 _____

5. 7,572 _____

6. 4,140 _____

7. 9,454 _____

8. 6,905 _____

Divide.

9. $15 \div 3 =$ _____

10. $49 \div 7 =$ _____

11. $63 \div 9 =$ _____

12. $8 \div 8 =$ _____

13. $30 \div 5 =$ _____

14. $48 \div 6 =$ _____

Line Plots

For 1–3, use the line plot at the right.

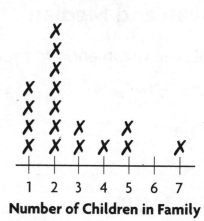

Number of Children in Family

1. The **✗**'s on this line plot represent the number of students. What do the numbers on the line plot represent?

   _____

2. What is the range of numbers used in this line plot?

   _____

3. What is the mode, or number that occurs most often, for this set of data?

   _____

4. Use the data in the table to complete the line plot.

Slices of Pizza Eaten	
Number of Slices	**Number of Students**
0	//
1	⊬⊬⊬ /
2	⊬⊬⊬
3	///
4	/
5	//

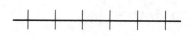

Slices of Pizza Eaten

Mixed Review

Find each product or quotient.

5. $10 \times 7 =$ ____   6. $7 \times 9 =$ ____   7. $6 \times 1 =$ ____   8. $8 \times 2 =$ ____

9. $8 \div 4 =$ ____   10. $36 \div 6 =$ ____   11. $0 \div 22 =$ ____   12. $45 \div 9 =$ ____

Mean and Median

Find the mean and the median.

1. 6, 3, 6

mean: _____

median: _____

2. 3, 8, 7

mean: _____

median: _____

3. 7, 2, 4, 8, 9

mean: _____

median: _____

4. 3, 4, 6, 4, 3

mean: _____

median: _____

5. 2, 8, 6, 5, 4, 3, 7

mean: _____

median: _____

6. 9, 2, 6, 3, 4, 1, 3

mean: _____

median: _____

Mixed Review

Write each missing number.

7. $6 + 0 =$ _____

8. $5 + 9 = 9 +$ _____

9. _____ $+ 8 = 8$

10. $2 \times$ _____ $= 2$

11. $4 \times$ _____ $= 0$

12. $3 \times 6 =$ _____ $\times 3$

13. $2 + (5 + 7) = (2 +$ _____ $) + 7$

14. $3 \times (4 + 2) = (3 \times$ _____ $) + (3 \times$ _____ $)$

15. $5 \times 8 = (5 \times 2) + (5 \times$ _____ $)$ **16.** $(4 \times 5) \times 2 = 4 \times ($ _____ $\times 2)$

List the coins and bills you would get as change from a $10 bill.

17. $9.43 _____

_____

18. $7.89 _____

_____

19. $4.10 _____

_____

20. $0.26 _____

_____

Problem Solving Strategy

Make a Graph

Choose one of the ideas shown
at the right for making a pictograph.

Take a survey to collect the data.
Then make a pictograph in the
space below. Decide on a symbol
and key for the graph. Include a
title and labels.

Pictograph—Menu of Ideas

Favorite Team Sport

Favorite Pizza Topping

Favorite TV Show

Key: Each _____ = _____ .

1. Tell how you chose a symbol, or picture, for your pictograph.

   _____

   _____

2. Explain how you chose a key for your pictograph.

   _____

   _____

Mixed Review

Write the value of the underlined digit.

3. 2,235 _____   4. 21,507 _____   5. 16,110 _____

Bar Graphs

For 1–4, use the bar graph.

1. What type of bar graph is this?

   _____

2. How many students named
   lions as their favorite stuffed
   animal? frogs? dogs?

   _____

   _____

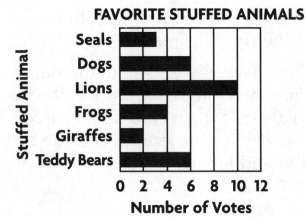

FAVORITE STUFFED ANIMALS

3. Which stuffed animal is the
   favorite of the most students?
   of the fewest students?

   _____

4. How many students in all
   voted for their favorite stuffed
   animal?

   _____

Mixed Review

Find the missing factor.

5. $20 = 10 \times$ _____   6. _____ $\times 3 = 27$ 7. $8 \times$ _____ $= 32$

8. _____ $\times 5 = 25$ 9. $6 \times$ _____ $= 24$ 10. $1 \times$ _____ $= 11$

11. $7 \times$ _____ $= 56$ 12. $24 = 8 \times$ _____   13. _____ $\times 6 = 0$

Solve.

14. $12 \div 2 =$ _____   15. $7 \div 1 =$ _____   16. $8 \div 2 =$ _____

17. $9 \div 3 =$ _____   18. $10 \div 5 =$ _____   19. $6 \div 3 =$ _____

20. $9 \times 9 =$ _____   21. $6 \times 9 =$ _____   22. $4 \times 7 =$ _____

23. 6,890
    +8,054

24. 3,211
    +7,618

25. 5,765
    +5,765

26. 9,298
    +5,431

Make Bar Graphs

Make a horizontal bar graph of the data in the table at the right. Use a scale of 2. Remember to write a title and labels for the graph.

FAVORITE DRINKS	
Drink	**Number of Votes**
Water	4
Punch	2
Milk	5
Juice	8
Soda	12

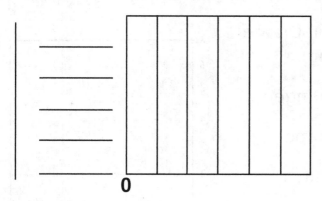

0

For 1–2, use your bar graph.

1. What does the graph show? _____

2. How many bars end halfway between two lines?

_____

Mixed Review

Write <, >, or = in each ◯.

3. 32 ÷ 8 ◯ 1 × 4   **4.** 6 + 6 ◯ 20   **5.** 5 × 2 ◯ 10 − 1

6. 7 × 7 ◯ 9 × 6   **7.** 18 ÷ 2 ◯ 3 + 11   **8.** 72 − 30 ◯ 9 × 3

© Harcourt

Algebra: Ordered Pairs

For 1–4, use the grid at the right. Write the letter of the point named by the ordered pair.

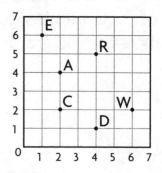

1. (4,5) _____   2. (1,6) _____

3. (6,2) _____   4. (2,2) _____

For 5–10, locate each ordered pair. Draw a point. Label it with the type of fruit.

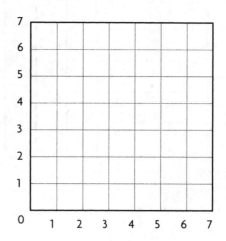

5. (1,1) apple   6. (5,5) orange

7. (2,4) banana   8. (3,2) grape

9. (4,3) kiwi   10. (6,1) peach

Mixed Review

Find the missing factor.

11. $3 \times$ _____ $= 21$ 12. $4 \times$ _____ $= 16$ 13. _____ $\times 4 = 24$

14. $7 \times$ _____ $= 56$ 15. _____ $\times 9 = 54$ 16. $5 \times$ _____ $= 50$

Solve.

17. 767
    -234

18. 9,870
    $-5,925$

19. 611
    $+382$

20. 2,195
    $+8,214$

21. $0 \times 8 =$ ___   22. $3 \times 5 =$ ___   23. $48 \div 8 =$ ___   24. $81 \div 9 =$ ___

25. $2 \times 10 =$ ___   26. $9 \times 8 =$ ___   27. $36 \div 4 =$ ___   28. $42 \div 7 =$ ___

29. $4 \times 3 =$ ___   30. $5 \times 6 =$ ___   31. $12 \div 1 =$ ___   32. $0 \div 7 =$ ___

Line Graphs

For 1–5, use the line graph at the right.

1. Joyce made this line graph to show the number of pages she read each day in a mystery book. On what day did Joyce read the most pages? the fewest?

_____

2. How many pages did Joyce read on Thursday?

_____

3. On which two days did Joyce read the same number of pages?

_____

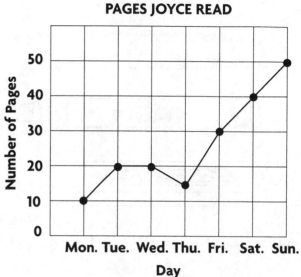

PAGES JOYCE READ

4. How many more pages did Joyce read on Friday than on Monday?

_____

5. Describe a trend you see on the line graph.

_____

Mixed Review

Solve.

6. $3\overline{)18}$ 7. $5\overline{)25}$ 8. $6\overline{)24}$ 9. $7\overline{)63}$

10. $10\overline{)10}$ 11. $8\overline{)24}$ 12. $10\overline{)20}$ 13. $2\overline{)14}$

14. $\begin{array}{r} 1{,}234 \\ +5{,}673 \end{array}$ 15. $\begin{array}{r} 3{,}179 \\ +3{,}298 \end{array}$ 16. $\begin{array}{r} 2{,}051 \\ -1{,}009 \end{array}$ 17. $\begin{array}{r} 8{,}233 \\ -4{,}649 \end{array}$

Length

Estimate the length in inches. Then use a ruler to measure to the nearest inch.

		Estimate	Measure

1. 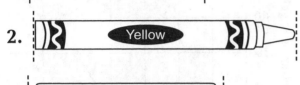 _____ _____

2. 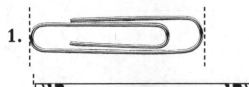 _____ _____

3.  _____ _____

Measure the length to the nearest half inch.

4.  _____ _____

5.  _____ _____

6.  _____ _____

Mixed Review

Solve.

7. $8 \times 6 =$ _____   8. $4 \times$ _____ $= 36$ 9. $72 =$ _____ $\times 9$

10. $7 \times 7 =$ _____   11. $3 \times$ _____ $= 21$ 12. $6 \times 9 =$ _____

Find the mean and the median.

13. 3, 8, 4   14. 2, 9, 7   15. 5, 6, 1

mean: _____   mean: _____   mean: _____

median: _____   median: _____   median: _____

Inch, Foot, Yard, and Mile

Choose the unit you would use to measure each.
Write *inch, foot, yard,* or *mile.*

1. the length of a table

_____

2. the length of a pine cone

_____

3. the length of a driveway

_____

4. the distance to the next town

_____

Choose the best unit of measure. Write *inches, feet,*
yards, or *miles.*

5. A pencil is about

5 _____ long.

6. The distance from your home
to the library is about 2

_____.

7. A bike is about

4 _____ long.

8. The woman bought about

4 _____ of fabric.

9. A sports card is about

3 _____ long.

10. A man is about

6 _____ tall.

Mixed Review

Find each product.

11. $7 \times 2 =$ _____

12. _____ $= 9 \times 5$

13. $6 \times 6 =$ _____

Find each quotient.

14. $14 \div 2 =$ _____

15. $27 \div 3 =$ _____

16. _____ $= 18 \div 6$

17. $24 \div 6 =$ _____

18. _____ $= 20 \div 4$

19. $8 \div 4 =$ _____

Capacity

Circle the better estimate.

1.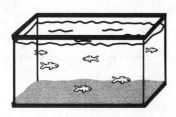

10 quarts or 10 gallons

2.

2 cups or 2 quarts

Compare. Write <, >, or = in each ◯.

3. 3 cups ◯ 1 pint

4. 1 gallon ◯ 4 quarts

5. 3 pints ◯ 2 quarts

6. 1 gallon ◯ 10 cups

7. 7 pints ◯ 1 gallon

8. 2 gallons ◯ 16 pints

Mixed Review

9. $\begin{array}{r} 6 \\ \times 8 \\ \hline \end{array}$

10. $\begin{array}{r} 9 \\ \times 9 \\ \hline \end{array}$

11. $\begin{array}{r} 86 \\ - 51 \\ \hline \end{array}$

12. $\begin{array}{r} 99 \\ - 83 \\ \hline \end{array}$

13. $7\overline{)63}$

14. $5\overline{)40}$

15. $6\overline{)24}$

16. $1\overline{)12}$

17. Find the sum of 862 and 137.

_____

18. Find the product of 6 and 9.

_____

19. Which number is greater: 736 or 763.

_____

20. Which number is less: 432 or 423.

_____

21. What is 4×8?

_____

22. What is $64 \div 8$?

_____

Weight

Choose the unit you would use to weigh each.
Write *ounce* or *pound*.

1.

_____

2.

_____

3.

_____

4.

_____

5.

_____

6.

_____

Circle the better estimate.

7.

4 ounces or
4 pounds

8.

10 ounces or
10 pounds

9.

10 ounces or
10 pounds

Mixed Review

Write the numbers in order from least to greatest.

10. 234, 561, 144 _____

11. 899, 998, 989 _____

12. 1,482; 1,248; 1,842 _____

13. 6,479; 8,372; 8,362 _____

Write the missing factor.

14. $4 \times$ _____ $= 16$ **15.** $12 = 6 \times$ _____   **16.** $3 \times$ _____ $= 27$

17. $80 =$ _____ $\times 8$ **18.** _____ $\times 3 = 30$ **19.** $487 =$ _____ $\times 487$

Ways to Change Units

Table of Measures	
Length	Capacity
12 inches = 1 foot	2 pints = 1 quart
3 feet = 1 yard	4 quarts = 1 gallon

Complete. Use the Table of Measures to help.

1. Change yards to feet.

larger unit _____

1 yard = _____

2. Change gallons to quarts.

larger unit _____

1 gallon = _____

Change the units. Use the Table of Measures to help.

3. _____ pints = 1 quart

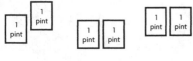

_____ pints = 5 quarts

4. _____ inches = 1 foot

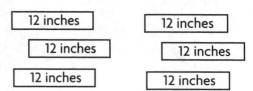

_____ inches = 6 feet

5. _____ cups = 1 quart

cups	4	8	12	16
quarts	1	2	3	4

_____ cups = 3 quarts

6. _____ feet = 1 yard

feet	3	6	9	12
yards	1	2	3	4

_____ feet = 4 yards

Mixed Review

Multiply.

7. $8 \times 9 =$ _____

8. $10 \times 4 =$ _____

9. $6 \times 7 =$ _____

Divide.

10. $18 \div 9 =$ _____

11. $36 \div 4 =$ _____

12. $40 \div 8 =$ _____

© Harcourt

Problem Solving Skill

Estimate or Measure

SUPPLIES	
Item	**Size**
Fabric	5 yd piece
Fabric	10 yd piece
Fringe	4 ft roll
Fringe	3 yd roll

For 1–2, use the table. Tell if you need to measure or if an estimate will do.

1. The Art Club is making 6 banners. For each banner, 4 feet of fabric is needed. Which piece of fabric should the Art Club buy? Explain.

_____

_____

2. For each of the 6 banners, 2 feet of fringe is needed. What rolls of fringe should the Art club buy? Explain.

_____

_____

Mrs. Winters is buying rice to cook for a family dinner. Rice is packaged in 2-pound bags. She needs to make enough rice to serve 12 people.

3. For each serving, 4 ounces of rice are needed. How many bags of rice should Mrs. Winters buy?

 A 2   C 4
 B 3   D 6

4. Which tool can Mrs. Winters use to measure the rice for one serving?

 F ruler   H balance
 G yardstick   J thermometer

Mixed Review

5. $\begin{array}{r} 359 \\ + 264 \\ \hline \end{array}$

6. $\begin{array}{r} 7,826 \\ + 1,358 \\ \hline \end{array}$

7. $\begin{array}{r} 213 \\ - 156 \\ \hline \end{array}$

8. $\begin{array}{r} 4,000 \\ - 2,479 \\ \hline \end{array}$

Write <, >, or = in the ◯.

9. 326 ◯ 362   10. 4,973 ◯ 4,973   11. 17,824 ◯ 17,631

Length

Estimate the length in centimeters. Then use a ruler to measure to the nearest centimeter.

1.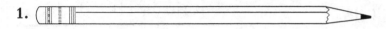

_____

2.

3. Yellow

_____   _____

Choose the unit you would use to measure each.
Write *cm, m,* or *km*.

4. the length of your
   little finger

_____

5. the distance between 2 towns

_____

6. the length of a chalkboard

_____

7. the length of your math book

_____

8. the length of the Mississippi
   River

_____

9. the length of a poster

_____

Mixed Review

10. $3.68
    − $1.79
    _____

11. 752
    + 134
    _____

12. 54 ÷ _____ = 6

13. 8 × 0 = _____

14. 5 ÷ _____ = 5

15. 7 × _____ = 56

Find the pattern and solve.

16. 64, 56, 48, 40, 32, _____

17. 1, 3, 5, 7, 9, 11, _____

18. 12, 18, 24, 30, _____, 42

19. 37, 34, 31, 28, _____, 22

Problem Solving Strategy

Make a Table

Complete this table.

1.

Meters	1	2	3	4	5	6	7	8	9	10
Centimeters	100	200								

For 2–3, use the completed table above.

2. Gary needs 500 centimeters of space for a bookcase. How many meters of space does he need?

_____

3. Kara needs 9 meters of string. How many centimeters of string does she need?

_____

Jake drew a line that was 3 decimeters long. How many centimeters long was his line?

4. Which table helps solve the problem? _____

A

Kilometers	1	2	3
Meters	1,000	2,000	3,000

C

Centimeters	100	200	300
Meters	1	2	3

B

Meters	1	2	3
Decimeters	10	20	30

D

Decimeters	1	2	3
Centimeters	10	20	30

5. What is the solution to the problem? _____

Mixed Review

Draw the next 3 shapes in the pattern.

6. ▽ ☐ △ ☐ ▽ ☐ △ ☐ ▽ ☐ △ ☐ _____

7. ○ ○ ○ ☐ ○ ○ ○ ☐ ○ ○ ☐ _____

Capacity

Circle the better estimate.

1.

1 mL or 1 L

2.

4 mL or 4 L

3.

15 mL or 15 L

4.

250 mL or 250 L

5.

2 mL or 2 L

6.

3,000 mL or 3,000 L

Choose the unit you would use to measure each.
Write *mL* or *L*.

7. a mug of hot
chocolate

_____

8. water in a
swimming pool

_____

9. a glass of juice

_____

10. water for a flower
garden

_____

11. a can of soup

_____

12. 5 pitchers of
lemonade

_____

Mixed Review

13. 59 + 64 + 93 = _____

14. 726 − 493 = _____

Write <, >, or = in each ◯.

15. 7 × 8 ◯ 87 − 31

16. 56 ÷ 7 ◯ 3 × 2

17. 40 ÷ 8 ◯ 7

18. 9 × 2 ◯ 6 × 3

Continue each pattern.

19. 8, 16, 24, 32, _____

20. 4, 9, 14, 19, _____, _____

21. 2, 5, 8, 11, _____, 17

22. 17, 15, 13, 11, _____

Name _____

Mass

Circle the better estimate.

1.

6 g or 6 kg

2.

25 g or 25 kg

3.

22 g or 22 kg

4.

4 g or 4 kg

5.

6 g or 6 kg

6.

2 g or 2 kg

Choose the tool and unit to measure each.

Tools	Units	
ruler	cm	g
liter container	kg	mL
simple balance	L	m

7. the mass of
a computer disk

_____

8. the length of
a desk

_____

9. the capacity of
a sink

_____

10. the mass of a
sack of sugar

_____

11. the length of
your hand

_____

12. the mass of
two bricks

_____

13. the mass of
a feather

_____

14. the mass of
an eraser

_____

Mixed Review

Solve.

15. $36 \div$ _____ $= 9$

16. _____ $\times 6 = 54$

17. $4 \times$ _____ $= 28$

18. _____ $\div 3 = 4$

19. $428 - 375 =$ _____

20. $32 + 69 + 51 =$ _____

21. $8 \times 0 =$ _____

22. $10 \div 1 =$ _____

Fahrenheit and Celsius

Write each temperature in °F.

1.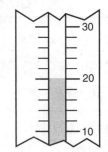

2.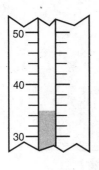

3.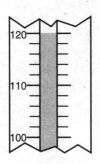

4.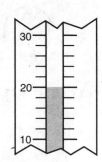

_____   _____   _____   _____

Write each temperature in °C.

5.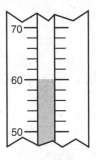

6.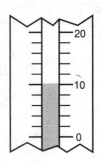

7.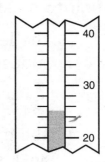

8.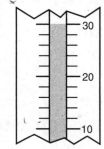

_____   _____   _____   _____

Circle the better estimate.

9.

10.

11.

12.

40°C or 0°C   5°C or 90°C   85°F or 32°F   5°F or 65°F

Mixed Review

Write $<$, $>$, or $=$ in each ◯.

13. $70 \div 7$ ◯ 11

14. $34 + 48$ ◯ 76

15. 42 ◯ 5×9

16. 8×3 ◯ 21

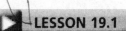

Line Segments and Angles

Name each figure.

1.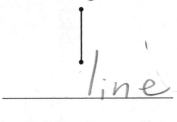

   line

2.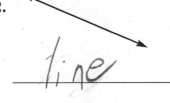

   line

3.

   angle

4.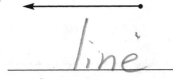

   line

5.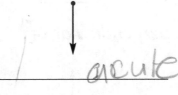

   acute

6.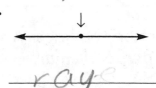

   ray

Use a corner of a sheet of paper to tell whether each
angle is a *right angle,* an *acute angle,* or an *obtuse angle.*

7.

   rite angle

8.

   acute

9.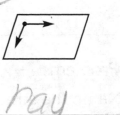

   ray

Draw each figure. You may wish to use a ruler or
straightedge.

10. line

11. ray

12. acute angle

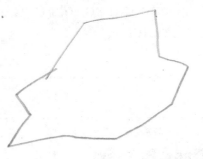

Mixed Review

Find each product.

13. 7
    ×6
    ____
    38

14. 5
    ×9
    ____
    45

15. 8
    ×8
    ____
    13

16. 4
    ×7
    ____
    28

Write <, >, or = in each ◯.

17. 8 + 9 ◯ 8 × 9

18. 24 + 16 + 52 ◯ 10 × 9

Name _____

Types of Lines

Describe the lines. Write *parallel* or *intersecting*.

 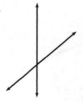 

1. _____ 2. _____ 3. _____

Describe the lines. Write *perpendicular* or
not perpendicular.

 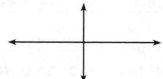 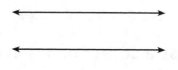

4. _____ 5. _____ 6. _____

For Problems 7–9, use the map at the right.

7. Name the streets that intersect
   Winter Street.

   _____

   _____

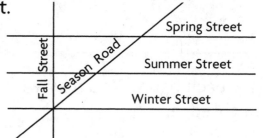

8. Name the streets that are
   parallel.

   _____

   _____

9. Name a road that is not
   perpendicular to Fall Street.

   _____

Mixed Review

Solve.

10. $5 \times 9 =$ _____   11. $7 \times 0 =$ _____

12. $4 \times 7 =$ _____   13. $6 \times 6 =$ _____

14. $27 \div 3 =$ _____   15. $32 \div 8 =$ _____

Plane Figures

Tell if each figure is a polygon. Write *yes* or *no*.

1.

2.

3.

4.

5.

_____ _____ _____ _____ _____

Write the number of sides and angles each polygon has.
Then name the polygon.

6.

7.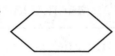

8.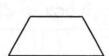

9.

_____ _____ _____ _____

_____ _____ _____ _____

_____ _____ _____ _____

10.

11.

12.

13.

_____ _____ _____ _____

_____ _____ _____ _____

_____ _____ _____ _____

Mixed Review

Decide if the number sentence is true or false. Write *true* or *false*.

14. $18 - 6 \neq 12$

15. $14 + 3 \neq 27$

16. $7 \times 6 = 42$

_____ _____ _____

17. $18 \div 6 \neq 2$

18. $5 \times 7 = 12$

19. $36 \div 6 = 6$

_____ _____ _____

Write $+$, $-$, $\div$, or $\times$ in the $\bigcirc$ to make the number sentence true.

20. $11 \bigcirc 8 = 19$

21. $24 \bigcirc 8 = 3$

22. $9 \bigcirc 9 = 81$

23. $35 \bigcirc 5 = 30$

24. $11 \bigcirc 7 = 77$

25. $42 \bigcirc 21 = 21$

© Harcourt

Name _____

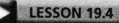

Triangles

For 1–3, use the triangles at the right. Write *A*, *B*, or *C*.

1. Which triangle is scalene? _____

2. Which triangles have at least
   2 equal sides? _____

3. Which triangle has 1 angle that
   is greater than a right angle? _____

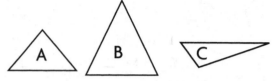

Write one letter from each box to describe each triangle.

a. equilateral triangle	**d.** right triangle
b. isosceles triangle	**e.** obtuse triangle
c. scalene triangle	**f.** acute triangle

4.
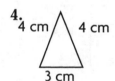

5.
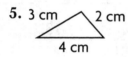

6.
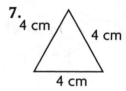

7.
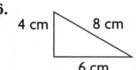

_____   _____   _____   _____

Name each triangle. Write *equilateral, isosceles,* or *scalene*. Then write
right, obtuse, or *acute.*

8.
4 cm
4 cm   4 cm

9.

10.
5 cm   3 cm
4 cm

11.
6 cm
2 cm
5 cm

_____   _____   _____   _____

_____   _____   _____   _____

Mixed Review

12.	4,692	13.	9,721	14.	6,400	15.	4,209
	+ 8,403		+ 3,688		+ 7,211		+ 362

Name _____

Quadrilaterals

For 1–3, use the quadrilaterals below. Write A, B, C, D, or E.

1. Which quadrilaterals have 2 pairs of equal sides? _____

2. Which quadrilaterals have no right angles? _____

3. How are quadrilaterals A and B alike? How are they different?

_____

_____

_____

For 4–7, write *all* the letters that describe each quadrilateral.
Then write a name for each quadrilateral.

a. It has 4 equal sides.   c. It has 4 right angles.

b. It has 2 pairs of parallel sides.   d. It has 2 pairs of equal sides.

4. 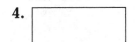   5.   6.   7.

_____ _____ _____ _____

_____ _____ _____ _____

Mixed Review

8. $3 + 3 + 3 + 3 + 3 + 3 =$ _____   9. $7 + 7 + 7 + 7 + 7 + 7 =$ _____

Describe the lines. Write *intersecting* or *parallel*.

10.   11.    12.

_____   _____   _____

Which number is less?

13. 4,375 or 4,735   14. 1,002 or 854   15. 2,014 or 2,004

_____   _____   _____

Problem Solving Strategy

Draw a Diagram

Describe where each figure should be in
the Venn diagram. Explain.

Isosceles Triangles   Right Triangles

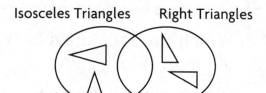

1. 3 cm 6 cm 5 cm

_____

_____

2. 4 cm 6 cm 4 cm

_____

_____

3. 5 cm 5 cm 4 cm

_____

_____

Mixed Review

Write whether each angle is a *right angle*, *acute angle*, or
obtuse angle.

4.   **5.**    **6.**

_____   _____   _____

Solve.

7.   352
     + 498

8.   1,867
     + 5,394

9.   841
     − 269

10.   403
      − 114

11.   4,306
      + 7,997

12.   9,294
      − 7,358

13.   6,845
      + 8,736

14.   7,000
      − 3,259

Congruent Figures

Fill in the blank.

1. _____ figures have the same size and shape.

2. Compare figures A and B.
   Are the figures congruent?
   Explain.

   | A | B |

   _____

   _____

   _____

3. Compare figures C and D.
   Are the figures congruent?
   Explain.

   _____

   _____

   _____

Trace and cut out each pair of figures. Tell if the figures
are congruent. Write *yes* or *no*.

4.

   _____

5.

   _____

6.

   _____

7.

   _____

8.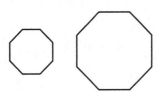

   _____

9.

   _____

Mixed Review

Solve.

10. Write the following numbers
    from least to greatest.
    384, 356, 383, 365

    _____

11. Marc buys a book for $5.95.
    He gives the clerk $10.00.
    What is his change?

    _____

Symmetry

Draw the line or lines of symmetry.

1.

2.

3.

Tell if the dashed line is a line of symmetry. Write
yes or *no*.

4.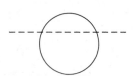

_____

5.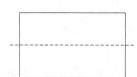

_____

6.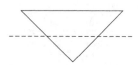

_____

7.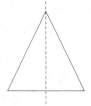

_____

8.

_____

9.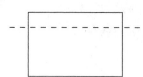

_____

Mixed Review

Solve.

10. Steve had 24 baseball cards.
He gave 10 cards to his sister.
Then he divided the rest of
the cards evenly between his
2 brothers. How many cards
did each brother get?

_____

11. Jem shared a package of 50
stickers equally among herself
and 4 friends. How many
stickers did each person
receive?

_____

Similar Figures

Fill in the blank.

1. Figures that have the same shape but may have

   different sizes are called _____ figures.

For 2–3, draw a similar figure. Use the grids below.

2.

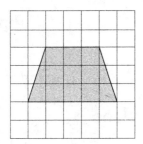

3.

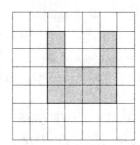

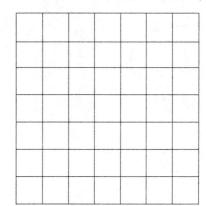

4. Draw a figure on one grid.
   Then draw a similar figure
   on the other grid.

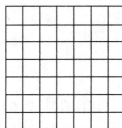

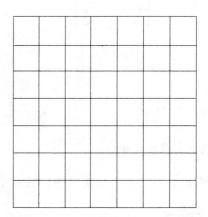

Mixed Review

Find the product.

5. $(2 \times 5) \times 6 =$ _____   6. $(1 \times 7) \times 8 =$ _____   7. $4 \times (3 \times 3) =$ _____

Slides, Flips, and Turns

Fill in the blank.

1. You _____ a figure when you move it in a straight line.

2. You _____ a figure when you move it over a line.

3. You _____ a figure when you rotate it around a point.

> slide
>
> turn
>
> flip

Tell what kind of motion was used to move each plane figure. Write *slide, flip,* or *turn.*

4.

_____

5.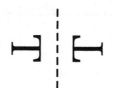

_____

6.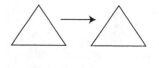

_____

7.

_____

8.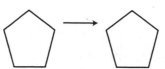

_____

9.

_____

Mixed Review

Predict the next three numbers in each pattern.

10. 20, 24, 28, 32, _____, _____, _____

11. 45, 39, 33, 27, _____, _____, _____

12. 33, 30, 27, 24, _____, _____, _____

13. 12, 16, 20, 24, _____, _____, _____

© Harcourt

Problem Solving Strategy

Make a Model

Make a model to solve.

1. Jeff made the trapezoid shown at the right with pattern blocks. What is another combination of pattern blocks that can be used to make a trapezoid that is congruent to this one?

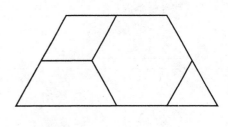

2. Use a different combination of pattern blocks to make another congruent trapezoid.

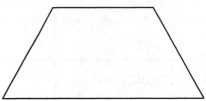

For 3–4, use pattern blocks to solve.

3. How many green triangles are needed to make a figure that is congruent to a yellow hexagon?

   A 3   **C** 6

   B 4   **D** 8

4. How many blue rhombuses are needed to make a figure that is congruent to a yellow hexagon?

   F 2   **H** 4

   G 3   **J** 5

Mixed Review

Tell if each figure is a polygon. Write *yes* or *no*.

5.

6.

7.

_____   _____   _____

Name _____

Solid Figures

Name the solid figure that each object looks like.

1.

_____

2.

_____

3.

_____

4.

_____

5.

_____

6.

_____

Complete the table.

	Figure	Faces	Edges	Vertices
7.	Cube			
8.	Rectangular prism			
9.	Square pyramid			

Mixed Review

Compare the numbers. Write <, >, or = in each ◯.

10. 3,535 ◯ 3,355

11. 67,100 ◯ 67,010

12. 53,701 ◯ 53,701

13. 9,999 ◯ 10,000

Find the quotient.

14. $25 \div 5 =$ ____

15. $45 \div 9 =$ ____

16. $35 \div 7 =$ ____

17. $50 \div 10 =$ ____

18. $49 \div 7 =$ ____

19. $15 \div 5 =$ ____

20. $81 \div 9 =$ ____

21. $54 \div 6 =$ ____

Find the difference.

22. $25 - 5 =$ ____

23. $45 - 9 =$ ____

24. $35 - 7 =$ ____

25. $50 - 10 =$ ____

26. $49 - 7 =$ ____

27. $15 - 5 =$ ____

28. $81 - 9 =$ ____

29. $54 - 6 =$ ____

© Harcourt

Name _____

Combine Solid Figures

Name the solid figures used to make each object.

1.

_____

2.

_____

3.

_____

4.

_____

5.

_____

6.

_____

Each pair of objects should be the same. Name the solid figure that is missing.

7.

_____

8.
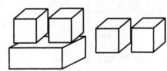

_____

9.
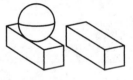

_____

10.
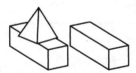

_____

11.
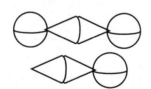

_____

12.
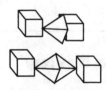

_____

Mixed Review

Round to the nearest ten.

13. 431 _____

14. 7,897 _____

15. 25,005 _____

16. 19,999 _____

Write the value of the underlined digit.

17. 1,298

18. 10,118

19. 900,255

20. 243,611

_____   _____   _____   _____

Tessellations

Tell if each figure will tessellate. Write *yes* or *no*.

1.

2.

3.

4.

_____   _____   _____   _____

Trace and cut out each figure. Use each figure to make a tessellation.
You may color your design.

5.

6.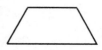

Mixed Review

Write each number in standard form.

7. 20,000 + 800 + 5   8. 30,000 + 6,000 + 10   9. 50,000 + 7,000 + 3

_____   _____   _____

Estimate each sum.

10.	874	11.	952	12.	892	13.	352	14.	925
	+ 635		+ 411		+ 999		+ 429		+ 659

Write the number of sides and angles each plane figure has.

15. hexagon   16. octagon   17. pentagon

_____   _____   _____

© Harcourt

Draw Figures

Write the number of line segments needed to draw each figure.

1. square _____   2. pentagon _____   3. trapezoid _____

Copy the solid figure. Name the figure.

4.

_____

Draw the missing line segments so that each figure matches its label.

5.

hexagon

6.

parallelogram

7.

octagon

8. Trace the figure shown at the right. Cut out the figure along the solid lines. Then fold along the dotted lines. Tape the edges of the figure together. What solid figure do you have?

_____

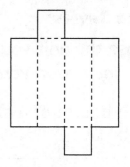

Mixed Review

Find the missing factor.

9. $7 \times$ _____ $= 21$ 10. _____ $\times 4 = 4$ 11. $6 \times$ _____ $= 48$

12. $9 \times$ _____ $= 0$ 13. _____ $\times 7 = 35$ 14. _____ $\times 9 = 63$

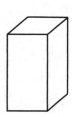

Problem Solving Skill
Identify Relationships

1. Look at the side of this rectangular prism.
   What plane figure describes the shape you see?

_____

For 2–5, use the figures below.

Figure K　　**Figure L**　　**Figure M**　　**Figure N**

2. Which figure is the side view
   of a square pyramid?

   A Figure K　　**C** Figure M
   B Figure L　　**D** Figure N

3. Which figure is the side view
   of a cube?

   F Figure K　　**H** Figure M
   G Figure L　　**J** Figure N

4. Which figure is the top view
   of a sphere?

   A Figure K　　**C** Figure M
   B Figure L　　**D** Figure N

5. Which figure is the bottom
   view of a cylinder?

   F Figure K　　**H** Figure M
   G Figure L　　**J** Figure N

Mixed Review

Choose the unit you would use to measure each.
Write *inch*, *foot*, *yard*, or *mile*.

6. the height of a chair

   _____

7. the length of a river

   _____

8. the length of your arm

   _____

9. the length of your classroom

   _____

Name _____

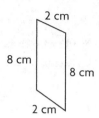

Perimeter

Find the perimeter.

1.

2.
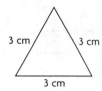

3.
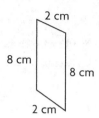

_____ units

_____

_____

Estimate the perimeter in centimeters. Then use your centimeter ruler to find the perimeter.

4.
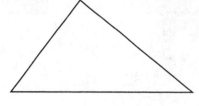

_____

5.

_____

6.
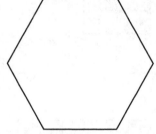

_____

7.

_____

Mixed Review

Use the graph.

8. How many students chose blue as their favorite color?

_____

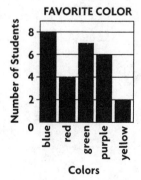

FAVORITE COLOR

9. How many more students chose green than yellow?

10. How many students voted in all?

_____

_____

Area

Find the area of each figure. Write the area in square units.

1.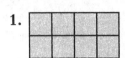

_____

2.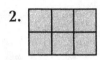

_____

3.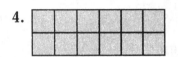

_____

4.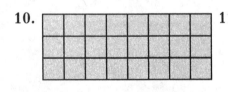

_____

5.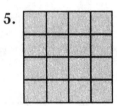

_____

6.

_____

7.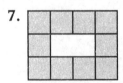

_____

8.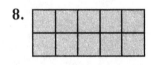

_____

9.

_____

10.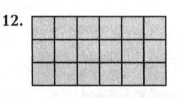

_____

11.

_____

12.

_____

Mixed Review

Find each missing number.

13. $4 + \underline{\hspace{0.5cm}} = 11$

14. $5 + \underline{\hspace{0.5cm}} = 8$

15. $9 + \underline{\hspace{0.5cm}} = 17$

16. $2 + \underline{\hspace{0.5cm}} = 10$

17. $\underline{\hspace{0.5cm}} \times 8 = 64$

18. $\underline{\hspace{0.5cm}} \times 8 = 32$

Problem Solving Skill

Make Generalizations

1. A laundry room is shaped like a rectangle. The area of the room is 6 square yards. The perimeter is 10 yards. The room is longer than it is wide. How wide is the room? How long is the room?

2. Mark has a piece of string that is 12 inches long. He shapes the string into a rectangle that has an area of 5 square inches. Can Mark make a shape that has a greater area with the string? If so, what is the area?

3. The perimeter of a table is 24 feet. The table is twice as long as it is wide. How long and how wide is the table?

4. Mrs. Brown put a wallpaper border around a room that is 10 feet long and 9 feet wide. How long is the wallpaper border? What is the area of the floor in the room?

Mixed Review

Solve.

5. The time shown on Mario's watch is 10:45. He has just finished raking leaves for 30 minutes. Before that, he played basketball for 1 hour. At what time did he start playing basketball?

6. Carrie is swimming in the middle lane of the pool. She waves to her father, who is swimming 3 lanes away, in the end lane. How many lanes does the pool have?

7. $\begin{array}{r} 9 \\ \times 6 \\ \hline \end{array}$
8. $\begin{array}{r} 5 \\ \times 7 \\ \hline \end{array}$
9. $\begin{array}{r} 7 \\ \times 7 \\ \hline \end{array}$
10. $\begin{array}{r} 8 \\ \times 3 \\ \hline \end{array}$
11. $\begin{array}{r} 4 \\ \times 6 \\ \hline \end{array}$

Volume

Use cubes to make each solid. Then write the volume in cubic units.

1.

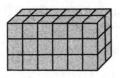

volume: _____

2.

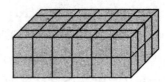

volume: _____

3.

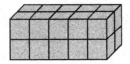

volume: _____

4.

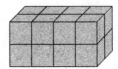

volume: _____

Find the volume of each solid. Write the volume in cubic units.

5.

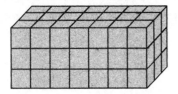

volume: _____

6.

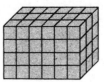

volume: _____

7.

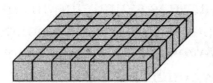

volume: _____

8.

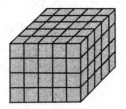

volume: _____

Mixed Review

Add.

9. 532
   $+\ 196$

10. 158
   $+\ 270$

11. 851
   $+\ 653$

12. 936
   $+\ 498$

Geometric Patterns

Name the pattern unit for each.

1. △ ○ ○ △ ○ ○ △ ○ ○ △ ○ ○

_____

2.

_____

Draw the next two shapes in each pattern.

3.  _____ _____

4. △ ⬯ ○ △ ⬯ ○ △ ⬯ ○ _____ _____

5. ◺ ⬜ ○ ◺ ⬜ ○ ◺ ⬜ ○ _____ _____

6. Jason used letters to make this pattern.

A B A B A B

How could he make the same pattern using pattern blocks?

_____

_____

7. Rita drew the pattern below. Show her pattern using the letters A and B.

_____

Mixed Review

8. $5 \times 7 =$ _____   9. $9 \times 6 =$ _____   10. $7 \times 8 =$ _____

Visual Patterns

Write a rule for each pattern.

_____

_____

_____

For 4–5, use the tile pattern below.

4. What is a rule for the pattern?

_____

5. Describe the next figure in the pattern.

_____

Mixed Review

6. $36 \div 9 =$ _____   **7.** $42 \div 6 =$ _____   **8.** $72 \div 8 =$ _____

Number Patterns

Write a rule for each pattern.

1. 29, 31, 33, 35, 37, 39   2. 87, 82, 77, 72, 67, 62

_____   _____

3. 350, 342, 334, 326, 318, 310   4. 491, 511, 531, 551, 571, 591

_____   _____

Write a rule for each pattern. Then find the missing numbers.

5. 67, 63, 59, 55, _____, 47, _____, _____

_____

6. 15, 24, 33, 42, _____, _____, 69, _____

_____

7. 592, 595, 598, 601, _____, 607, _____, _____

_____

8. 726, 711, 696, 681, _____, _____, 636, _____

_____

Mixed Review

Write each number in standard form.

9. three thousand, seven _____

10. fifty thousand, nine hundred forty-two _____

11. seventy-one thousand, sixty _____

12. two hundred four thousand, one hundred eight _____

Solve.

13.	14.	15.	16.
2,308	7,416	3,957	6,004
+5,897	−4,329	+7,168	−4,836

Make Patterns

For 1–2, use these shapes.

1. Choose 3 shapes. Use them to draw a pattern unit.
   Repeat the pattern unit two times.

2. Choose 4 shapes. Use them to draw a pattern unit.
   Repeat the pattern unit two times.

For 3–4, you may wish to use a calculator.

3. Think of a 2-digit number. Write a rule for a pattern so
   that a 1-digit number is added to find the next number.
   Write the first 4 numbers in the pattern.

   Rule: _____

   _____ , _____ , _____ , _____

4. Think of a 3-digit number. Write a rule for a pattern
   so that a 2-digit number is subtracted to find the next
   number. Write the first 4 numbers in the pattern.

   Rule: _____

   _____ , _____ , _____ , _____

Mixed Review

Write the time shown on each clock.

5.

6.

7.

_____   _____   _____

Problem Solving Strategy

Find a Pattern

Use *find a pattern* to solve.

The map shows how some houses on Bay Avenue are
numbered. Use the map for 1–4.

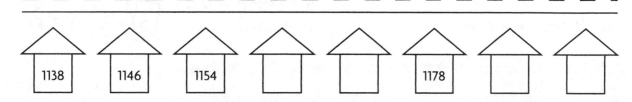

1138   1146   1154   1178

1. Michael is trying to deliver
   a pizza to the family at 1186
   Bay Avenue. To which house
   should he deliver it?

   _____

2. What rule describes how the
   houses are numbered?

   _____

3. What are the numbers of the
   fourth and fifth houses?

   _____

4. A new house is being built to
   the right of 1194 Bay Avenue.
   What should its address be?

   _____

Mixed Review

Add.

5.	6.	7.	8.
257	463	1,289	$23.48
132	288	7,411	$19.95
+319	+504	+3,926	+$44.50

Write $<$, $>$, or $=$ for each $\bigcirc$.

9. 3,295 $\bigcirc$ 3,259   10. 64,086 $\bigcirc$ 64,086   11. 15,107 $\bigcirc$ 15,017

Find each missing factor.

12. $4 \times$ _____ $= 32$ 13. $9 \times$ _____ $= 63$ 14. _____ $\times 5 = 30$

15. $7 \times$ _____ $= 49$ 16. _____ $\times 3 = 24$ 17. _____ $\times 6 = 48$

Probability

Tell whether each event is *certain* or *impossible*.

1. Pencils will fall from the sky.

_____

2. Winter in Alaska will be cold.

_____

For 3–4, look at the cards. Suppose they are mixed up and placed face-down. You turn over one card.

3. Name the number you are most unlikely to choose. _____

4. Name the number you are most likely to choose. _____

For 5–8, look at the numbered tile at the right. Write *impossible, unlikely, likely,* or *certain* to match each event.

1	3	3
1	5	7
3	5	5

5. dropping a coin on an odd number

_____

6. dropping a coin on a 7

_____

7. dropping a coin on a number greater than 9

_____

8. dropping a coin on a number less than 6

_____

Ryan tossed two cubes each numbered 1 through 6. He adds the numbers.

9. Is it likely or unlikely that he will get a sum less than 12?

_____

10. The first number he rolls is 3. Is it certain or impossible that he will get a sum greater than 9?

_____

Mixed Review

Find the product.

11. $9 \times 8 =$ _____   **12.** $7 \times 6 =$ _____   **13.** $6 \times 4 =$ _____   **14.** $5 \times 9 =$ _____

● **Outcomes**

For 1–4, list the possible outcomes of each experiment.

1. dropping a marker on one of these squares

1		
3	11	
5	7	9

_____

2. pulling a number from this bag

_____

3. tossing a cube labeled A–F

_____

4. using this spinner

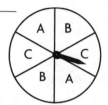

_____

5. Karen has a bag of 4 blue balls, 2 green balls, and 1 red ball. What is the chance that she will pull a green ball from the bag?

_____

6. Martin spins the pointer. What is his chance of spinning a square?

_____

7. Gia used this spinner. The pointer landed on black 1 time, and on white 1 time. Predict the color it will land on next. What is the chance she will spin gray?

_____

Mixed Review

Round to the nearest thousand.

8. 2,780 _____

9. 1,376 _____

10. 4,900 _____

11. 3,100 _____

Find the missing addend.

12. 900 + _____ = 1,000   13. _____ + 779 = 979   14. 954 + _____ = 1,250

Experiments

Read the following experiment.

Marsha has a bag filled with 20 tiles. There are 7 blue,
2 green, 4 yellow, and 7 red tiles. She pulls a tile from the bag
10 times. Below is a list of the outcomes of the 10 pulls.

1–red	6–red
2–blue	7–blue
3–red	8–yellow
4–yellow	9–red
5–green	10–blue

Record the results in the tally table.

MARSHA'S EXPERIMENT	
Color	Tally
Red	
Blue	
Yellow	
Green	

Use your tally table to answer 1–3.

1. What color did she pull most
   often?

   _____

2. What color did she pull least
   often?

   _____

3. Why do you think this is so?

   _____

   _____

   _____

Mixed Review

Solve.

4. 33
   +17
   ____

5. 79
   +82
   ____

6. 543
   +108
   ____

7. 412
   +344
   ____

8. 190
   +150
   ____

9. 222
   +279
   ____

10. 987
    +213
    ____

11. 557
    +904
    ____

12. $10 \times 4 =$ _____

13. _____ $\times\ 9 = 27$

14. $5 \times$ _____ $= 40$

Predict Outcomes

1. This tally table shows the pulls from a bag of tiles. Predict which color is most likely to be pulled.

Tally Table	
Color	**Tallies**
black	~~HHt~~
green	////
red	~~HHt~~ ~~HHt~~ //

_____

2. The line plot below shows the results of rolling a number cube. Predict which number you would most likely roll.

```
X  X  X  X  X  X
X  X  X  X  X  X
X  X  X  X  X  X
X  X  X  X  X  X
X  X  X  X  X  X
X  X  X  X  X  X
┼──┼──┼──┼──┼──┼
1  2  3  4  5  6
```

_____

3. This tally table shows the results of using a spinner. Predict whether the pointer will land on blue or red on the next spin.

Tally Table	
Color	**Tallies**
blue	~~HHt~~ ~~HHt~~ ~~HHt~~ /
red	~~HHt~~ ~~HHt~~ ~~HHt~~ /

_____

4. This tally table shows the pulls from a bag of balls. Predict which color is least likely to be pulled.

Tally Table	
Color	**Tallies**
blue	~~HHt~~ ~~HHt~~ ~~HHt~~ ////
white	//
purple	~~HHt~~ ~~HHt~~ ///

_____

Mixed Review

Complete.

5. 35¢ = _____ pennies

6. $2.00 = _____ dimes

7. 75¢ = _____ quarters

8. 65¢ = _____ nickels

Underline the number that is less.

9. 35 or 54   **10.** 91 or 88   **11.** 110 or 100

Combinations

For 1–2, make a tree diagram to show all the
combinations. Tell how many combinations are possible.

1. Pizza Crusts: Thin, Thick
   Pizza Toppings:
       Cheese, Sausage, Pepperoni

2. Colors of Poster Board:
       White, Pink, Yellow
   Colors of Letters:
       Red, Black, Blue, Green

_____ combinations

_____ combinations

3. Amy is buying a new car. She
   can choose a sports car, a
   mid-size car, or a sports utility
   vehicle. The color choices are
   white, black, or silver. Draw a
   tree diagram that shows the
   possible combinations of car
   types and colors.

_____ combinations

4. A store sells four different colors of sweatshirts in three different
   sizes. Write a multiplication sentence to find the total number of
   color and size combinations.

_____

Mixed Review

5. $9\overline{)81}$

6. $5\overline{)10}$

7. $6\overline{)36}$

8. $7\overline{)49}$

9. 9×3

10. 7×6

11. 4×8

12. 6×6

Problem Solving Strategy

Make an Organized List

Make a list to solve. If you need more space, write your
list on the back of your paper.

1. How many ways can you
   arrange the numbers 2, 4, 6,
   and 8 in a 4-digit number?

   _____

   _____

   _____

2. How many ways can you
   arrange turkey, cheese,
   lettuce, and tomato on a
   sandwich?

   _____

   _____

   _____

   _____

   _____

   _____

   _____

   _____

   _____

3. How many ways can you order
   math, science, and language
   arts homework?

   _____

   _____

   _____

4. How many ways can Martin
   order his calls to Emily, Kim,
   and Ben on the phone?

   _____

   _____

   _____

Mixed Review

Find the sum or the difference.

5. $\begin{array}{r} 7{,}594 \\ +3{,}928 \\ \hline \end{array}$

6. $\begin{array}{r} 4{,}766 \\ -3{,}827 \\ \hline \end{array}$

7. $\begin{array}{r} \$43.57 \\ +\$26.84 \\ \hline \end{array}$

8. $\begin{array}{r} \$25.89 \\ -\ \$6.92 \\ \hline \end{array}$

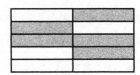

Parts of a Whole

Write a fraction in numbers and in words that names the shaded part.

1.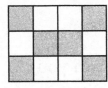

_____

2.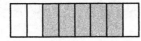

_____

3.

_____

Write the fraction, using numbers.

4. three fifths

_____

5. six out of ten

_____

6. two divided by three

_____

7. one out of six

_____

8. nine divided by ten

_____

9. seven twelfths

_____

Write a fraction to describe the part of each figure that is shaded.

10.   _____

Mixed Review

Find the difference.

11. $85 - 29 =$ _____

12. $346 - 173 =$ _____

13. $811 - 559 =$ _____

14. $300 - 101 =$ _____

15. $924 - 474 =$ _____

16. $865 - 239 =$ _____

Find the product.

17. $0 \times 1 =$ _____

18. $3 \times 6 =$ _____

19. $10 \times 6 =$ _____

20. $8 \times 2 =$ _____

21. $7 \times 8 =$ _____

22. $5 \times 5 =$ _____

Name _____

Parts of a Group

Use a pattern to complete the table.

1.	Model	○ ○ ○	● ○ ○	● ● ○	
2.	Total number of parts	3		3	3
3.	Number of shaded parts		1	2	3
4.	Fraction of shaded parts	$\frac{0}{3}$	$\frac{1}{3}$		$\frac{3}{3}$

Write a fraction that names the part of each group that is circled.

5.

6.

7.

_____   _____   _____

8.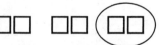

9.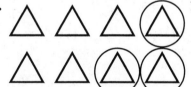

10.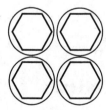

_____   _____   _____

Mixed Review

Find the quotient.

11. $6 \div 6 =$ _____

12. $0 \div 9 =$ _____

13. $5 \div 1 =$ _____

14. $16 \div 4 =$ _____

15. $10 \div 1 =$ _____

16. $12 \div 3 =$ _____

17. $28 \div 7 =$ _____

18. $30 \div 3 =$ _____

19. $16 \div 2 =$ _____

20. $64 \div 8 =$ _____

21. $42 \div 7 =$ _____

22. $72 \div 9 =$ _____

Equivalent Fractions

Find an equivalent fraction. Use fraction bars.

1.

$\frac{1}{3}$

_____

2.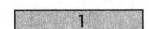

$\frac{1}{8}$ $\frac{1}{8}$ $\frac{1}{8}$ $\frac{1}{8}$ $\frac{1}{8}$ $\frac{1}{8}$

_____

3.

1

$\frac{1}{8}$ $\frac{1}{8}$ $\frac{1}{8}$ $\frac{1}{8}$

_____

4.

1

$\frac{1}{6}$ $\frac{1}{6}$ $\frac{1}{6}$ $\frac{1}{6}$

_____

5.

$\frac{1}{10}$ $\frac{1}{10}$

_____

6.

1

$\frac{1}{5}$ $\frac{1}{5}$ $\frac{1}{5}$ $\frac{1}{5}$

_____

Find the missing numerator. Use fraction bars.

7. $\frac{1}{3} = \frac{\square}{6}$

8. $\frac{3}{5} = \frac{\square}{10}$

9. $\frac{3}{4} = \frac{\square}{8}$

10. $\frac{1}{5} = \frac{\square}{10}$

11. $\frac{12}{12} = \frac{\square}{6}$

12. $\frac{2}{3} = \frac{\square}{12}$

13. $\frac{6}{8} = \frac{\square}{4}$

14. $\frac{4}{5} = \frac{\square}{10}$

15. $\frac{2}{3} = \frac{\square}{6}$

16. $\frac{4}{8} = \frac{\square}{4}$

17. $\frac{3}{5} = \frac{\square}{10}$

18. $\frac{2}{12} = \frac{\square}{6}$

Mixed Review

Round to the nearest thousand.

19. 554 _____

20. 3,764 _____

21. 7,298 _____

22. 9,099 _____

Find the quotient.

23. $12 \div 3 =$ _____

24. $16 \div 8 =$ _____

25. $30 \div 3 =$ _____

26. $64 \div 8 =$ _____

27. $63 \div 7 =$ _____

28. $10 \div 1 =$ _____

29. $0 \div 6 =$ _____

30. $25 \div 5 =$ _____

31. $72 \div 8 =$ _____

32. $32 \div 4 =$ _____

33. $45 \div 5 =$ _____

34. $48 \div 6 =$ _____

Compare and Order Fractions

Compare. Write $<$, $>$, or $=$ in each ◯.

1.

$\frac{2}{3}$ ◯ $\frac{3}{6}$

2.

$\frac{3}{4}$ ◯ $\frac{4}{6}$

3.

$\frac{3}{5}$ ◯ $\frac{3}{4}$

4.

$\frac{4}{8}$ ◯ $\frac{1}{2}$

Compare the part of each group that is shaded.
Write $<$ or $>$ in each ◯.

5.

$\frac{3}{4}$ ◯ $\frac{1}{4}$

6.

$\frac{5}{8}$ ◯ $\frac{7}{8}$

Use fraction bars to compare.

7. Order $\frac{1}{2}$, $\frac{2}{3}$, and $\frac{3}{4}$ from greatest to least.

_____

8. Order $\frac{1}{8}$, $\frac{1}{3}$, and $\frac{3}{6}$ from greatest to least.

_____

Mixed Review

Compare. Write $<$, $>$, or $=$ in each ◯.

9. 472 ◯ 619   **10.** 3,009 ◯ 2,588   **11.** 820 ◯ 820

Order each set of numbers from least to greatest.

12. 35, 63, 17   **13.** 200, 199, 205   **14.** 484, 848, 488

_____   _____   _____

Problem Solving Strategy

Make A Model

Use *make a model* to solve.

1. Sean spent $\frac{2}{10}$ of his allowance on a book and $\frac{2}{5}$ on a baseball. On which item did he spend more?

2. Alex read $\frac{3}{8}$ of a book. Joel read $\frac{1}{2}$ of the same book. Who read more?

3. Mr. Ruiz made a divider for his patio. He used 9 stacks of bricks with 7 bricks in each stack. How many bricks did he use?

4. The border in Shea's room repeats square, triangle, triangle, circle. If one wall has 9 repeats, how many triangles are on that wall?

Mixed Review

5. Tia, Juan, and Carla are standing in a line. Tia is behind Juan. Carla is in front of Juan. In what order are they standing?

6. There are 67 marbles in a jar. Ed takes out 22 marbles on Monday. On Tuesday, Ed puts 35 marbles into the jar. How many marbles are in the jar now?

Complete.

7. 3 feet = __?__ yard

8. 1 foot = __?__ inches

9. 24, 26, 28, 30, _____

10. 14, 17, 20, 23, _____

Name _____

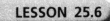

LESSON 25.6

Mixed Numbers

Write a mixed number for the parts that are shaded.

1. ☐☐☐

2. ☐☐☐☐

3.

_____   _____   _____

4. ☐☐

5. ☐☐

6.

_____   _____   _____

Use the number line to write the mixed number.

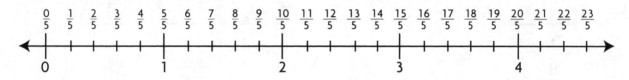

7. $\dfrac{6}{5}$ _____

8. $\dfrac{12}{5}$ _____

9. $\dfrac{16}{5}$ _____

10. $\dfrac{9}{5}$ _____

11. $\dfrac{17}{5}$ _____

12. $\dfrac{22}{5}$ _____

13. $\dfrac{7}{5}$ _____

14. $\dfrac{14}{5}$ _____

15. $\dfrac{20}{5}$ _____

Mixed Review

Find the product.

16. $3 \times 4 =$ _____

17. $5 \times 9 =$ _____

18. $4 \times 8 =$ _____

19. $4 \times 6 =$ _____

20. $3 \times 5 =$ _____

21. $9 \times 6 =$ _____

22. $2 \times 9 =$ _____

23. $7 \times 8 =$ _____

24. $3 \times 7 =$ _____

© Harcourt

Practice   PW135

Add Fractions

Find the sum.

1.

| $\frac{1}{4}$ | | $\frac{1}{4}$ |

$\frac{1}{4} + \frac{1}{4} =$ _____

2.

| $\frac{1}{6}$ | $\frac{1}{6}$ | $\frac{1}{6}$ | | $\frac{1}{6}$ |

$\frac{3}{6} + \frac{1}{6} =$ _____

3.

| $\frac{1}{5}$ | $\frac{1}{5}$ | $\frac{1}{5}$ | | $\frac{1}{5}$ |

$\frac{3}{5} + \frac{1}{5} =$ _____

4.

| $\frac{1}{8}$ | $\frac{1}{8}$ | | $\frac{1}{8}$ | $\frac{1}{8}$ | $\frac{1}{8}$ |

$\frac{2}{8} + \frac{3}{8} =$ _____

5.

| $\frac{1}{6}$ | $\frac{1}{6}$ | $\frac{1}{6}$ | $\frac{1}{6}$ | | $\frac{1}{6}$ |

$\frac{4}{6} + \frac{1}{6} =$ _____

6.

| $\frac{1}{4}$ | | $\frac{1}{4}$ | | $\frac{1}{4}$ |

$\frac{2}{4} + \frac{1}{4} =$ _____

Use fraction bars to find the sum.

7. $\frac{1}{10} + \frac{2}{10} =$ _____

8. $\frac{4}{10} + \frac{3}{10} =$ _____

9. $\frac{3}{5} + \frac{1}{5} =$ _____

10. $\frac{1}{4} + \frac{3}{4} =$ _____

11. $\frac{2}{5} + \frac{1}{5} =$ _____

12. $\frac{7}{12} + \frac{2}{12} =$ _____

Mixed Review

Add.

13. $3 + 4 + 5 =$ _____

14. $1 + 1 + 9 =$ _____

15. $5 + 8 + 7 =$ _____

Which is greater?

16. 5 feet or 5 inches

17. 2 feet or 2 yards

18. 6 cups or 6 pints

_____

_____

_____

Use fraction bars. Compare. Write <, >, or = in each $\bigcirc$.

19. $\frac{3}{5} \bigcirc \frac{1}{4}$

20. $\frac{2}{3} \bigcirc \frac{4}{6}$

21. $\frac{1}{8} \bigcirc \frac{2}{4}$

22. $\frac{5}{6} \bigcirc \frac{9}{10}$

23. $\frac{1}{2} \bigcirc \frac{1}{8}$

24. $\frac{2}{5} \bigcirc \frac{3}{4}$

Add Fractions

Find the sum. Write the answer in simplest form.

1.

| $\frac{1}{8}$ | $\frac{1}{8}$ | $\frac{1}{8}$ | $\frac{1}{8}$ | $\frac{1}{8}$ | $\frac{1}{8}$ |

| $\frac{1}{4}$ | $\frac{1}{4}$ | $\frac{1}{4}$ |

$\dfrac{4}{8} + \dfrac{2}{8} =$ _____

2.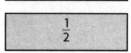

| $\frac{1}{12}$ | $\frac{1}{12}$ | $\frac{1}{12}$ | $\frac{1}{12}$ | $\frac{1}{12}$ | $\frac{1}{12}$ |

| $\frac{1}{2}$ |

$\dfrac{2}{12} + \dfrac{4}{12} =$ _____

3.

| $\frac{1}{5}$ | $\frac{1}{5}$ | $\frac{1}{5}$ | $\frac{1}{5}$ |

$\dfrac{3}{5} + \dfrac{1}{5} =$ _____

4.

| $\frac{1}{6}$ | $\frac{1}{6}$ |

| $\frac{1}{3}$ |

$\dfrac{1}{6} + \dfrac{1}{6} =$ _____

Find the sum. Write the answer in simplest form.
Use fraction bars if you wish.

5. $\dfrac{1}{6} + \dfrac{3}{6} =$ _____

6. $\dfrac{4}{12} + \dfrac{3}{12} =$ _____

7. $\dfrac{3}{8} + \dfrac{3}{8} =$ _____

8. $\dfrac{1}{4} + \dfrac{1}{4} =$ _____

9. $\dfrac{4}{12} + \dfrac{4}{12} =$ _____

10. $\dfrac{1}{2} + \dfrac{1}{2} =$ _____

11. $\dfrac{1}{6} + \dfrac{1}{6} =$ _____

12. $\dfrac{1}{8} + \dfrac{1}{8} =$ _____

13. $\dfrac{1}{12} + \dfrac{1}{12} =$ _____

14. $\dfrac{1}{10} + \dfrac{1}{10} =$ _____

15. $\dfrac{1}{5} + \dfrac{1}{5} =$ _____

16. $\dfrac{3}{4} + \dfrac{1}{4} =$ _____

Mixed Review

Write a fraction to describe the part of each group that is shaded.

17. ▦ ▦ ▦ ☐

_____

18. ⬤ ⬤ ⬤
⬤ ⬤ ◯

_____

19. ▦ ▦ ▦ ▦ ▦
▦ ▦ ▦ ▦ ☐

_____

Write the quotient.

20. $30 \div 3 =$ _____

21. $64 \div 8 =$ _____

22. $28 \div 7 =$ _____

Subtract Fractions

Find the difference.

1.

$\frac{1}{4}$	$\frac{1}{4}$	$\frac{1}{4}$

$\frac{3}{4} - \frac{2}{4} =$ _____

2.

$\frac{1}{6}$	$\frac{1}{6}$	$\frac{1}{6}$	$\frac{1}{6}$

$\frac{4}{6} - \frac{1}{6} =$ _____

3.

$\frac{1}{8}$	$\frac{1}{8}$	$\frac{1}{8}$	$\frac{1}{8}$	$\frac{1}{8}$	$\frac{1}{8}$

$\frac{6}{8} - \frac{2}{8} =$ _____

4.

$\frac{1}{5}$	$\frac{1}{5}$	$\frac{1}{5}$	$\frac{1}{5}$

$\frac{4}{5} - \frac{3}{5} =$ _____

Use fraction bars to find the difference.

5. $\frac{6}{10} - \frac{1}{10} =$ _____

6. $\frac{4}{10} - \frac{3}{10} =$ _____

7. $\frac{3}{5} - \frac{1}{5} =$ _____

8. $\frac{5}{8} - \frac{3}{8} =$ _____

9. $\frac{4}{5} - \frac{2}{5} =$ _____

10. $\frac{7}{12} - \frac{2}{12} =$ _____

11. $\frac{2}{3} - \frac{1}{3} =$ _____

12. $\frac{8}{8} - \frac{3}{8} =$ _____

13. $\frac{3}{4} - \frac{2}{4} =$ _____

14. $\frac{4}{6} - \frac{1}{6} =$ _____

15. $\frac{11}{12} - \frac{4}{12} =$ _____

16. $\frac{5}{6} - \frac{4}{6} =$ _____

Mixed Review

Solve.

17. $5 + (4 + 1) =$ _____

18. $(1 + 1) + 9 =$ _____

19. $8 + (7 + 5) =$ _____

20.
$$\begin{array}{r} 712 \\ -\ 558 \\ \hline \end{array}$$

21.
$$\begin{array}{r} 450 \\ +\ 388 \\ \hline \end{array}$$

22.
$$\begin{array}{r} 917 \\ -\ 652 \\ \hline \end{array}$$

Write the place value of the 2 in each number.

23. 23,957

24. 43,289

25. 88,072

_____

_____

_____

Subtract Fractions

Compare. Find the difference. Write the answer in simplest form.

1.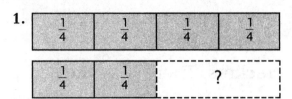

$$\frac{4}{4} - \frac{2}{4} = \underline{\hspace{1cm}}$$

2.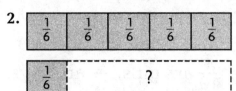

$$\frac{5}{6} - \frac{1}{6} = \underline{\hspace{1cm}}$$

3.

$$\frac{7}{8} - \frac{3}{8} = \underline{\hspace{1cm}}$$

4.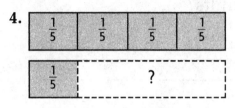

$$\frac{4}{5} - \frac{1}{5} = \underline{\hspace{1cm}}$$

Find the difference. Write the answer in simplest form.
Use fraction bars if you wish.

5. $\frac{6}{8} - \frac{2}{8} = \underline{\hspace{1cm}}$

6. $\frac{4}{10} - \frac{2}{10} = \underline{\hspace{1cm}}$

7. $\frac{4}{5} - \frac{1}{5} = \underline{\hspace{1cm}}$

8. $\frac{5}{8} - \frac{3}{8} = \underline{\hspace{1cm}}$

9. $\frac{4}{6} - \frac{2}{6} = \underline{\hspace{1cm}}$

10. $\frac{7}{12} - \frac{2}{12} = \underline{\hspace{1cm}}$

11. $\frac{5}{6} - \frac{1}{6} = \underline{\hspace{1cm}}$

12. $\frac{8}{8} - \frac{2}{8} = \underline{\hspace{1cm}}$

13. $\frac{6}{10} - \frac{2}{10} = \underline{\hspace{1cm}}$

14. $\frac{9}{10} - \frac{1}{10} = \underline{\hspace{1cm}}$

15. $\frac{11}{12} - \frac{2}{12} = \underline{\hspace{1cm}}$

16. $\frac{3}{4} - \frac{1}{4} = \underline{\hspace{1cm}}$

Mixed Review

Add.

17. $\frac{1}{4} + \frac{1}{4} = \underline{\hspace{1cm}}$

18. $\frac{1}{5} + \frac{3}{5} = \underline{\hspace{1cm}}$

19. $\frac{1}{6} + \frac{4}{6} = \underline{\hspace{1cm}}$

Complete.

20. $4 \times \underline{\hspace{1cm}} \times 3 = 12$

21. $5 \times \underline{\hspace{1cm}} \times 8 = 0$

22. $\underline{\hspace{1cm}} \times 8 \times 6 = 48$

Problem Solving Skill

Reasonable Answers

Solve. Tell how you know your answer is reasonable.

1. A table seats 10 people. Of the people sitting at the table, $\frac{4}{10}$ are girls, $\frac{4}{10}$ are boys, and the rest are adults. What part of the table is occupied by adults?

_____

_____

_____

_____

2. Perry opened a package of crackers. He ate $\frac{3}{8}$ of the crackers. Then Terry ate $\frac{2}{8}$ of the crackers. What part of the crackers were left?

_____

_____

_____

_____

3. Janet colored $\frac{7}{12}$ of her picture red and $\frac{3}{12}$ of her picture green. The rest of the picture was left uncolored. What part of her picture was left uncolored?

_____

_____

_____

_____

4. Michael opened a package of wrapping paper. He used $\frac{1}{4}$ of the paper to wrap a present and $\frac{1}{4}$ of the paper to decorate a box. How much of the paper was left?

_____

_____

_____

_____

Mixed Review

Solve.

5. $19 - 15 =$ _____

6. $72 \div 9 =$ _____

7. $39 - 27 =$ _____

Name _____

Fractions and Decimals

Write the fraction and decimal for the shaded part.

1. 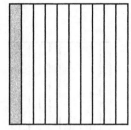 2. 3. 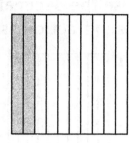 4.

_____   _____   _____   _____

5. 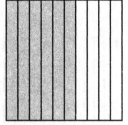 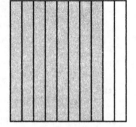 6. 7. 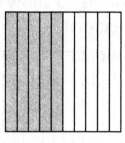 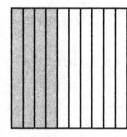 8.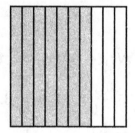

_____   _____   _____   _____

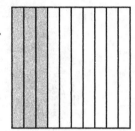

Mixed Review

Find the quotient.

9. $12 \div 2 =$ _____   10. $16 \div 8 =$ _____   11. $9 \div 3 =$ _____

12. $63 \div 9 =$ _____   13. $50 \div 10 =$ _____   14. $56 \div 7 =$ _____

15. $35 \div 5 =$ _____   16. $24 \div 4 =$ _____   17. $36 \div 4 =$ _____

Solve.

18. $\begin{array}{r} 484 \\ -232 \end{array}$ 19. $\begin{array}{r} 795 \\ +496 \end{array}$ 20. $\begin{array}{r} 734 \\ -207 \end{array}$ 21. $\begin{array}{r} 225 \\ +118 \end{array}$

22. $\begin{array}{r} 8,128 \\ -2,716 \end{array}$ 23. $\begin{array}{r} 4,030 \\ +1,812 \end{array}$ 24. $\begin{array}{r} 9,235 \\ -2,122 \end{array}$ 25. $\begin{array}{r} 5,687 \\ +3,401 \end{array}$

Tenths

Use the decimal models to show each amount. Then write
each fraction as a decimal.

1.
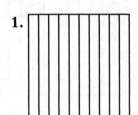

$\dfrac{2}{10}$ _____

2.
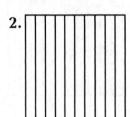

$\dfrac{9}{10}$ _____

3.
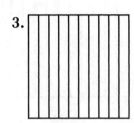

$\dfrac{3}{10}$ _____

4.
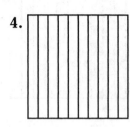

$\dfrac{1}{10}$ _____

Write each fraction as a decimal.

5. $\dfrac{4}{10}$ ____
6. $\dfrac{2}{10}$ ____
7. $\dfrac{1}{10}$ ____
8. $\dfrac{9}{10}$ ____
9. $\dfrac{7}{10}$ ____

Write each decimal as a fraction.

10. 0.5 ____
11. 0.3 ____
12. 0.8 ____
13. 0.6 ____
14. 0.9 ____

Mixed Review

Compare. Write $<$, $>$, or $=$ for each ◯.

15. 4×7 ◯ 5×5

16. 3×6 ◯ 9×2

17. 33 ◯ 4×8

18. 7×1 ◯ 14×0

19. 10×4 ◯ 47

20. 10×2 ◯ 5×4

Write each number in expanded form.

21. $32,594$ _____

22. $6,720$ _____

23. $40,897$ _____

24. $75,912$ _____

Name _____

Hundredths

Use the decimal models to show each amount. Then write
the decimal.

1.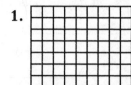
seven hundredths

_____

2.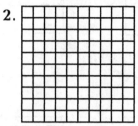
nine hundredths

_____

3.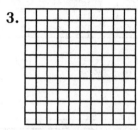
twenty hundredths

_____

4.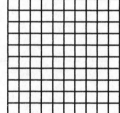
twenty-five hundredths

_____

5.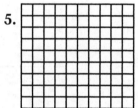
forty-nine hundredths

_____

6.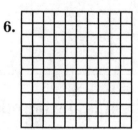
seventy-two hundredths

_____

Write each fraction as a decimal.

7. $\frac{25}{100}$ _____

8. $\frac{50}{100}$ _____

9. $\frac{85}{100}$ _____

10. $\frac{3}{100}$ _____

Write each decimal as a fraction.

11. 0.06 _____

12. 0.74 _____

13. 0.12 _____

14. 0.01 _____

Mixed Review

15. $5{,}591 + 332 =$ _____

16. $654 + 1{,}987 =$ _____

17. $3{,}069 + 451 =$ _____

18. $674 - 91 =$ _____

19. $274 - 115 =$ _____

20. $953 - 608 =$ _____

21. $4{,}124 - 1{,}325 =$ _____

22. $7{,}833 + 1{,}049 =$ _____

© Harcourt

Decimals Greater Than One

Write the word form and expanded form for each decimal.

1.
Ones	•	Tenths	Hundredths
1	•	2	7

_____

_____

2.
Ones	•	Tenths	Hundredths
3	•	9	1

_____

_____

3.
Ones	•	Tenths	Hundredths
5	•	4	5

_____

_____

4.
Ones	•	Tenths	Hundredths
7	•	6	8

_____

_____

Write the missing numbers.

5. $2.35 = 2 + \underline{\hspace{1cm}} + 0.05$

6. $6.79 = \underline{\hspace{1cm}} + 0.7 + \underline{\hspace{1cm}}$

7. $4.16 = 4 + \underline{\hspace{1cm}} + 0.06$

8. $2.51 = \underline{\hspace{1cm}} + 0.5 + \underline{\hspace{1cm}}$

Mixed Review

Find the product.

9. $4 \times 5 = \underline{\hspace{1cm}}$

10. $7 \times 9 = \underline{\hspace{1cm}}$

11. $6 \times 7 = \underline{\hspace{1cm}}$

12. $\underline{\hspace{1cm}} = 6 \times 6$

13. $5 \times 8 = \underline{\hspace{1cm}}$

14. $\underline{\hspace{1cm}} = 9 \times 3$

15. Kristi drinks 3 glasses of milk each day. How many glasses of milk does she drink in one week?

_____

16. A bus can seat 25 passengers. How many passengers can ride on 2 buses?

_____

Compare and Order Decimals

Compare. Write < or > for each ○.

1.
  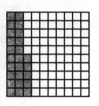

2.25 ○ 2.14

2.
Ones	•	Tenths	Hundredths
8	•	5	6
6	•	9	5

8.56 ○ 6.95

3.
Ones	•	Tenths	Hundredths
4	•	7	2
6	•	0	1

4.72 ○ 6.01

Use the number line to order the decimals from least to greatest.

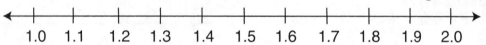
1.0 1.1 1.2 1.3 1.4 1.5 1.6 1.7 1.8 1.9 2.0

4. 1.6, 1.1, 1.9

_____

5. 1, 1.6, 1.1

_____

6. 1.3, 2.0, 1.6

_____

7. 1.9, 1, 2.0

_____

Mixed Review

Add. Write the answer in simplest form.

8. $\frac{1}{2} + \frac{1}{2} =$ _____

9. $\frac{1}{4} + \frac{1}{4} =$ _____

10. $\frac{2}{8} + \frac{3}{8} =$ _____

Subtract. Write the answer in simplest form.

11. $\frac{8}{10} - \frac{5}{10} =$ _____

12. $\frac{9}{12} - \frac{8}{12} =$ _____

13. $\frac{5}{6} - \frac{3}{6} =$ _____

Tell the time 3 hours after the time on each clock.

14. _____

15. _____

16. _____

© Harcourt

Problem Solving Skill

Too Much/Too Little Information

For 1–4, write *a*, *b*, or *c* to tell whether the problem has

a. too much
 information

b. too little
 information

c. the right
 amount of
 information

Solve those with too much or the right amount of information. Tell what is missing for those with too little information.

1. The vet weighed Jenny's cats. Mittens weighs 5.8 kilograms, Fluffy weighs 4.9 kilograms, and Boots weighs 5.6 kilograms. Which kitten weighs the least?

_____

_____

2. Richard bought a package of ground meat. He used it to make 6 equal-size burgers. How many ounces did each burger weigh?

_____

_____

3. Lisa bought 6 bread sticks and a pizza. The pizza was cut into 12 slices. She and her friends ate 9 slices of pizza. What fraction of the pizza was left?

_____

_____

4. Brady bought a pen that cost $1.24. He also bought a pencil. The clerk gave him $3.41 change. What was the cost of the pencil?

_____

_____

Mixed Review

Write the fraction that names the shaded part.

5. 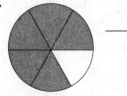 ___

6. 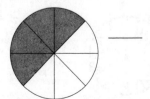 ___

7. 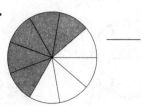 ___

© Harcourt

Name _____

Fractions and Money

Write the amount of money shown. Then write the
amount as a fraction of a dollar.

1.

_____

2.

_____

3.

_____

4.

_____

5.

_____

6.

_____

7.

_____

8.

_____

Mixed Review

Write a decimal to show what part of each decimal
square is shaded.

9.

_____

10.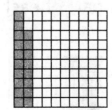

_____

11.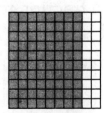

_____

Find the quotient.

12. $54 \div 9 =$ _____ **13.** $50 \div 5 =$ _____ **14.** $20 \div 5 =$ _____

© Harcourt

Decimals and Money

Write the money amount for each fraction of a dollar.

1. $\frac{20}{100}$ _____

2. $\frac{62}{100}$ _____

3. $\frac{25}{100}$ _____

4. $\frac{78}{100}$ _____

5. $\frac{55}{100}$ _____

6. $\frac{50}{100}$ _____

7. $\frac{15}{100}$ _____

8. $\frac{9}{100}$ _____

Write the money amount.

9. 32 hundredths of a dollar

_____

10. 9 hundredths of a dollar

_____

11. 48 hundredths of a dollar

_____

12. 99 hundredths of a dollar

_____

13. 61 hundredths of a dollar

_____

14. 5 hundredths of a dollar

_____

Write the missing numbers. Use the fewest coins possible.

15. $0.36 = _____ dimes _____ pennies

16. $0.05 = _____ dimes _____ pennies

17. $0.64 = _____ dimes _____ pennies

18. $0.14 = _____ dimes _____ pennies

Mixed Review

Write a fraction to show what part of each decimal model is shaded.

19.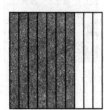

_____

20.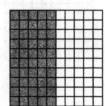

_____

21.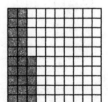

_____

Add and Subtract Decimals and Money

Add or subtract.

1. $\begin{array}{r} 0.27 \\ + 0.39 \\ \hline \end{array}$

2. $\begin{array}{r} 0.70 \\ - 0.16 \\ \hline \end{array}$

3. $\begin{array}{r} 0.88 \\ - 0.29 \\ \hline \end{array}$

4. $\begin{array}{r} 0.26 \\ + 0.35 \\ \hline \end{array}$

5. $\begin{array}{r} 0.47 \\ + 0.26 \\ \hline \end{array}$

6. $\begin{array}{r} 0.99 \\ - 0.37 \\ \hline \end{array}$

7. $\begin{array}{r} 0.31 \\ + 0.47 \\ \hline \end{array}$

8. $\begin{array}{r} 0.78 \\ - 0.46 \\ \hline \end{array}$

9. $\begin{array}{r} \$0.98 \\ -\$0.50 \\ \hline \end{array}$

10. $\begin{array}{r} \$0.58 \\ +\$0.21 \\ \hline \end{array}$

11. $\begin{array}{r} 0.81 \\ - 0.49 \\ \hline \end{array}$

12. $\begin{array}{r} 0.73 \\ + 0.12 \\ \hline \end{array}$

13. $\begin{array}{r} 1.00 \\ - 0.99 \\ \hline \end{array}$

14. $\begin{array}{r} 0.34 \\ + 0.56 \\ \hline \end{array}$

15. $\begin{array}{r} 0.89 \\ - 0.49 \\ \hline \end{array}$

16. $\begin{array}{r} 4.5 \\ + 3.6 \\ \hline \end{array}$

17. $\begin{array}{r} \$2.13 \\ + \$0.39 \\ \hline \end{array}$

18. $\begin{array}{r} \$4.89 \\ - \$2.37 \\ \hline \end{array}$

19. $\begin{array}{r} 0.18 \\ + 1.56 \\ \hline \end{array}$

20. $\begin{array}{r} 8.6 \\ - 3.9 \\ \hline \end{array}$

Mixed Review

Add or subtract.

21. $\begin{array}{r} 243 \\ 45 \\ + 82 \\ \hline \end{array}$

22. $\begin{array}{r} 116 \\ 267 \\ + 96 \\ \hline \end{array}$

23. $\begin{array}{r} 89 \\ 74 \\ + 96 \\ \hline \end{array}$

24. $\begin{array}{r} 741 \\ 249 \\ + 331 \\ \hline \end{array}$

25. $\begin{array}{r} 99 \\ - 55 \\ \hline \end{array}$

26. $\begin{array}{r} 965 \\ - 832 \\ \hline \end{array}$

27. $\begin{array}{r} 8,100 \\ - 4,900 \\ \hline \end{array}$

28. $\begin{array}{r} 78 \\ - 45 \\ \hline \end{array}$

© Harcourt

Problem Solving Strategy

Solve a Simpler Problem

Use the prices in the chart.
Solve a simpler problem.

1. Pam has $4. If she buys 1 box of crayons and 3 tubes of paint, how much money will she have left?

crayons	$0.39 per box
markers	$0.75 per box
paints	$0.85 per tube
brush	$0.28

_____

2. Stephano has $3. If he buys one of everything on the price list, how much money will he have left?

3. Daniel has $6. If he buys 3 tubes of paints and 3 brushes, how much money will he have left?

_____

Mixed Review

Write the amount of money shown. Then write the amount as a fraction of a dollar.

4.

5.

6.

_____   _____   _____

Find the perimeter of each figure.

7.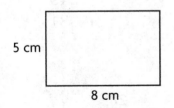
5 cm
8 cm

8.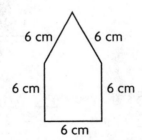
6 cm   6 cm
6 cm   6 cm
6 cm

9.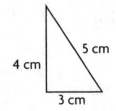
4 cm   5 cm
3 cm

_____   _____   _____

© Harcourt

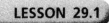

Algebra: Multiply Multiples of 10 and 100

Complete. Use patterns and mental math to help.

1. $9 \times 1 =$ _____   **2.** $6 \times 3 =$ _____

$9 \times 10 =$ _____ $6 \times 30 =$ _____

$9 \times 100 =$ _____ $6 \times 300 =$ _____

3. $7 \times 4 =$ _____   **4.** $6 \times 5 =$ _____

_____ $\times 40 = 280$ _____ $\times 50 = 300$

$7 \times$ _____ $= 2,800$ $6 \times$ _____ $= 3,000$

Use mental math and basic facts to complete.

5. $7 \times 80 =$ _____   **6.** $9 \times$ _____ $= 4,500$ **7.** _____ $\times 60 = 240$

8. $2 \times$ _____ $= 1,400$ **9.** $7 \times$ _____ $= 4,200$ **10.** _____ $\times 800 = 2,400$

11. _____ $\times 20 = 180$ **12.** $5 \times 500 =$ _____   **13.** $5 \times 400 =$ _____

14. $3 \times$ _____ $= 210$ **15.** $1 \times$ _____ $= 100$ **16.** $5 \times 200 =$ _____

Mixed Review

Add or subtract.

17. $\begin{array}{r} 3.5 \\ -\ 2.4 \\ \hline \end{array}$ **18.** $\begin{array}{r} 4.07 \\ +\ 3.72 \\ \hline \end{array}$ **19.** $\begin{array}{r} 5.8 \\ -\ 4.5 \\ \hline \end{array}$ **20.** $\begin{array}{r} 4.72 \\ -\ 2.31 \\ \hline \end{array}$ **21.** $\begin{array}{r} 2.3 \\ +\ 6.5 \\ \hline \end{array}$

Multiply or divide.

22. $36 \div 6 =$ _____   **23.** $18 \div 6 =$ _____   **24.** $10 \times 6 =$ _____

25. $81 \div 9 =$ _____   **26.** $7 \times 6 =$ _____   **27.** $56 \div 8 =$ _____

Multiply 2-Digit Numbers

Find the product.

1.

$2 \times 10 = 20$ $2 \times 4 = 8$

$2 \times 14 =$ _____

2.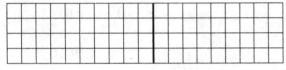

$3 \times 10 = 30$ $3 \times 2 = 6$

$3 \times 12 =$ _____

3.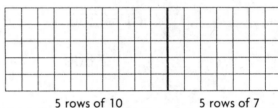

$4 \times 10 = 40$ $4 \times 3 = 12$

$4 \times 13 =$ _____

4.

4 rows of 10	4 rows of 9
$4 \times 10 = 40$	$4 \times 9 = 36$

$4 \times 19 =$ _____

5.

5 rows of 10	5 rows of 7
$5 \times 10 = 50$	$5 \times 7 = 35$

$5 \times 17 =$ _____

Find the product. You may wish to use base-ten blocks or grid paper.

6. $4 \times 12 =$ _____

7. $3 \times 13 =$ _____

Mixed Review

Add or subtract.

8.
$$\begin{array}{r} 62 \\ -\ 33 \\ \hline \end{array}$$

9.
$$\begin{array}{r} \$0.38 \\ +\ \$0.19 \\ \hline \end{array}$$

10.
$$\begin{array}{r} 79 \\ +\ 28 \\ \hline \end{array}$$

11.
$$\begin{array}{r} 54 \\ +\ 42 \\ \hline \end{array}$$

12.
$$\begin{array}{r} 94 \\ -\ 59 \\ \hline \end{array}$$

13.
$$\begin{array}{r} 88 \\ +\ 17 \\ \hline \end{array}$$

14.
$$\begin{array}{r} \$0.68 \\ -\ \$0.47 \\ \hline \end{array}$$

15.
$$\begin{array}{r} \$0.76 \\ -\ \$0.39 \\ \hline \end{array}$$

Find the product.

16.
$$\begin{array}{r} 8 \\ \times\ 7 \\ \hline \end{array}$$

17.
$$\begin{array}{r} 9 \\ \times\ 3 \\ \hline \end{array}$$

18.
$$\begin{array}{r} 10 \\ \times\ 5 \\ \hline \end{array}$$

19.
$$\begin{array}{r} 7 \\ \times\ 9 \\ \hline \end{array}$$

20.
$$\begin{array}{r} 6 \\ \times\ 9 \\ \hline \end{array}$$

Problem Solving Skill

Choose the Operation

Write whether you would *add*, *subtract*, *multiply*, or *divide*. Then solve.

1. Susan's family paid $36 for 4 videos. Each video cost the same amount. How much did each video cost?

   _____

2. A third-grade class learns 18 spelling words one week and 16 the next week. How many words does the class learn in 2 weeks?

   _____

3. The lunch room can seat 84 students. If there are 56 students in the lunch room, how many more students can the lunch room hold?

   _____

4. Maria has written 24 pages in her diary. She puts 3 daily entries on each page. How many daily entries has she written?

   _____

Mixed Review

Find the sum.

5. $\begin{array}{r} 14 \\ 15 \\ + 18 \end{array}$	6. $\begin{array}{r} 29 \\ 8 \\ + 77 \end{array}$	7. $\begin{array}{r} 63 \\ 30 \\ + 49 \end{array}$	8. $\begin{array}{r} 47 \\ 114 \\ + 142 \end{array}$	9. $\begin{array}{r} 20 \\ 67 \\ + 38 \end{array}$	10. $\begin{array}{r} 83 \\ 25 \\ + 71 \end{array}$
11. $\begin{array}{r} 753 \\ + 495 \end{array}$	12. $\begin{array}{r} 934 \\ + 248 \end{array}$	13. $\begin{array}{r} 295 \\ + 692 \end{array}$	14. $\begin{array}{r} 854 \\ + 196 \end{array}$	15. $\begin{array}{r} 717 \\ + 362 \end{array}$	
16. $\begin{array}{r} 4,762 \\ + 3,291 \end{array}$	17. $\begin{array}{r} 9,132 \\ + 4,376 \end{array}$	18. $\begin{array}{r} 5,689 \\ + 8,542 \end{array}$	19. $\begin{array}{r} 1,911 \\ + 8,149 \end{array}$	20. $\begin{array}{r} 7,571 \\ + 6,025 \end{array}$	
21. $\begin{array}{r} \$14.29 \\ + \$\ 6.33 \end{array}$	22. $\begin{array}{r} \$\ 4.10 \\ + \$27.19 \end{array}$	23. $\begin{array}{r} \$2.05 \\ + \$8.99 \end{array}$	24. $\begin{array}{r} \$62.77 \\ + \$18.19 \end{array}$	25. $\begin{array}{r} \$41.95 \\ + \$27.42 \end{array}$	

Choose a Method

Multiply.

| 1. 54
× 5 | 2. 26
× 3 | 3. 19
× 7 | 4. 29
× 2 | 5. 53
× 4 |

Find the product. Estimate to check.

| 6. 76
× 9 | 7. 95
× 7 | 8. 63
× 2 | 9. 38
× 3 | 10. 42
× 5 |

Find the product.

| 11. 48
× 6 | 12. 73
× 7 | 13. 37
× 9 | 14. 62
× 8 | 15. 25
× 7 |

| 16. 46
× 5 | 17. 83
× 8 | 18. 79
× 6 | 19. 93
× 5 | 20. 56
× 7 |

| 21. 58
× 4 | 22. 20
× 6 | 23. 87
× 3 | 24. 16
× 4 | 25. 32
× 5 |

Mixed Review

Write the time.

26.

27.

28.

_____   _____   _____

Divide with Remainders

Vocabulary

Fill in the blank.

1. In division, the _____ is the amount left over when a number cannot be divided evenly.

Use counters to find the quotient and remainder.

2. $13 \div 3 =$ _____   3. $15 \div 2 =$ _____   4. $11 \div 4 =$ _____

5. $12 \div 5 =$ _____   6. $10 \div 4 =$ _____   7. $9 \div 5 =$ _____

Find the quotient and remainder. You may use counters or draw a picture to help.

8. $17 \div 3 =$ _____   9. $13 \div 4 =$ _____

10. $23 \div 4 =$ _____   11. $30 \div 4 =$ _____

12. $25 \div 3 =$ _____   13. $17 \div 4 =$ _____

Mixed Review

Find the difference. Estimate to check.

14. $432 - 251 =$ 15. $847 - 563 =$ 16. $712 - 386 =$

_____   _____   _____

17. $598 - 202 =$ 18. $\$6.29 - \$3.84 =$ 19. $515 - 409 =$

_____   _____   _____

20. $\$7.06 - \$4.37 =$ 21. $824 - 399 =$ 22. $918 - 264 =$

_____   _____   _____

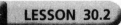

Divide 2-Digit Numbers

Use the model. Write the quotient and remainder.

1.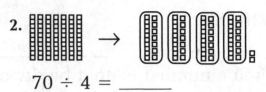

$51 \div 2 = $ _____.

2.

$70 \div 4 = $ _____

Divide. You may use base-ten blocks to help.

3. $61 \div 4 = $ _____

4. $17 \div 2 = $ _____

5. $63 \div 5 = $ _____

Divide and check.

6. $7\overline{)72}$

7. $6\overline{)49}$

8. $3\overline{)38}$

9. $5\overline{)59}$

10. $3\overline{)49}$

11. $8\overline{)87}$

Solve the division problem. Then write the *check* step.

	Check:	Check:	Check:
12. $5\overline{)27}$		13. $3\overline{)48}$	14. $4\overline{)65}$

Mixed Review

Find the product.

15. $\begin{array}{r} 13 \\ \times\ 6 \\ \hline \end{array}$

16. $\begin{array}{r} 21 \\ \times\ 3 \\ \hline \end{array}$

17. $\begin{array}{r} 53 \\ \times\ 5 \\ \hline \end{array}$

18. $\begin{array}{r} 36 \\ \times\ 4 \\ \hline \end{array}$

19. $\begin{array}{r} 19 \\ \times\ 1 \\ \hline \end{array}$

20. $\begin{array}{r} 48 \\ \times\ 7 \\ \hline \end{array}$

Problem Solving Skill

Interpret the Remainder

1. Alexandra has 74 baseball cards in a collection. She can fit 9 cards on a page. How many pages does she need?

_____

2. Roger is making kites. It takes 6 feet of string to make a kite. He has 80 feet of string. How many kites can he make?

_____

3. Clem has 63 books. He wants to put an equal number of books on each of 5 shelves. The rest of the books he will donate to a library. How many books will Clem donate to a library?

_____

4. George is making toast. His toaster toasts 2 slices of bread at one time. He cannot toast one slice at a time in his toaster. He has 19 pieces of bread. How many times will he use his toaster?

_____

5. Rob has 32 snacks. He needs to pack an equal number into each of 5 boxes. How many snacks will be in each box?

_____

6. Mary and 12 of her friends are going on a bus trip. Each seat on the bus holds three. How many seats will they need?

_____

Mixed Review

Divide and check.

7. $9\overline{)37}$

8. $8\overline{)46}$

9. $4\overline{)58}$

Subtract.

10. 4,236
    −3,572

11. 3,502
    −2,508

12. 4,003
    −3,927

13. 8,611
    −7,844

Name _____

Divide 3-Digit Numbers

Divide.

1. 5)810   2. 3)963   3. 6)948   4. 7)952

5. 4)392   6. 2)830   7. 7)924   8. 5)255

9. 2)174   10. 9)675   11. 8)744   12. 3)762

Mixed Review

Multiply.

13. 21
× 3

14. 76
× 8

15. 26
× 7

16. 21
× 4

17. 38
× 5

18. 25
× 2

19. 34
× 6

20. 52
× 9

Estimate Quotients

Estimate each quotient. Write the basic fact you used to
find the estimate.

1. $179 \div 3$ **2.** $484 \div 7$ **3.** $199 \div 4$

_____   _____   _____

4. $416 \div 6$ **5.** $648 \div 9$ **6.** $137 \div 2$

_____   _____   _____

Estimate the quotient.

7. $148 \div 5 = $ _____   **8.** $134 \div 7 = $ _____   **9.** $268 \div 3 = $ _____

10. $555 \div 7 = $ _____   **11.** $538 \div 9 = $ _____   **12.** $334 \div 8 = $ _____

13. $3 \overline{)142}$ **14.** $7 \overline{)500}$ **15.** $3 \overline{)299}$

16. $5 \overline{)444}$ **17.** $8 \overline{)317}$ **18.** $8 \overline{)635}$

Mixed Review

Divide.

19. $9 \overline{)36}$ **20.** $7 \overline{)49}$ **21.** $3 \overline{)15}$ **22.** $5 \overline{)45}$

23. $9 \overline{)81}$ **24.** $6 \overline{)54}$ **25.** $9 \overline{)54}$ **26.** $4 \overline{)32}$

Find the product.

27. $\begin{array}{r} 38 \\ \times\ 6 \\ \hline \end{array}$ **28.** $\begin{array}{r} 57 \\ \times\ 4 \\ \hline \end{array}$ **29.** $\begin{array}{r} 69 \\ \times\ 3 \\ \hline \end{array}$ **30.** $\begin{array}{r} 84 \\ \times\ 7 \\ \hline \end{array}$